A Laboratory Manual
for *Legionella*

A Laboratory Manual for *Legionella*

Edited by

T. G. Harrison and A. G. Taylor

A Wiley–Interscience Publication

JOHN WILEY AND SONS
Chichester · New York · Brisbane · Toronto · Singapore

Library of Congress Cataloging-in-Publication Data

A Laboratory manual for *Legionella* / edited by T.G. Harrison and A.G. Taylor.
p. cm.
'A Wiley–Interscience publication.'
Bibliography: p.
ISBN 0 471 91861 X
1. Legionnaires' disease—Diagnosis—Laboratory manuals.
2. *Legionella pneumophila*. 3. Diagnosis, Laboratory. I. Harrison, T. G. II. Taylor, A. G.
QR201.L44L33 1988
616.25—dc19 87—35094 CIP

British Library Cataloguing in Publication Data

A Laboratory manual for *Legionella*.
1. Legionnaires' disease
I. Harrison, T.G. II. Taylor, A.G.
616.2'41 RC152.7

ISBN 0 471 91861 X

Typeset by Woodfield Graphics, Arundel, West Sussex.
Printed in Great Britain.

Contents

Contributors

P. J. L. DENNIS
Quality Control and Safety Laboratory, PHLS Centre for Applied Microbiology and Research, Porton Down, Salisbury, Wiltshire SP4 0JG, UK

E. DOURNON
Laboratoire Central, Hôpital Claude Bernard, 10 Avenue de la Porte d'Aubervilliers, Paris 75019, France

T. G. HARRISON
Legionella Reference Unit, Division of Microbiological Reagents and Quality Control, Central Public Health Laboratory, 61 Colindale Avenue, London NW9 5HT, UK

C. MAYAUD
Service de Réanimation Pneumologique, Hôpital Tenon, 4 Rue de la Chine, Paris 75020, France

N. A. SAUNDERS
Legionella Reference Unit, Division of Microbiological Reagents and Quality Control, Central Public Health Laboratory, 61 Colindale Avenue, London NW9 5HT, UK

A. G. TAYLOR
Legionella Reference Unit, Division of Microbiological Reagents and Quality Control, Central Public Health Laboratory, 61 Colindale Avenue, London NW9 5HT, UK

R. WAIT
Bacterial Metabolism Research Laboratory, PHLS Centre for Applied Microbiology and Research, Porton Down, Salisbury, Wiltshire SP4 0JG, UK

Foreword

The dramatic outbreak of acute respiratory infections among the veterans attending a conference in Philadelphia in 1976, and the recognition of the new genus of bacteria – *Legionella* – responsible, have provided new diagnostic targets for medical microbiologists, and new sanitary targets for water and air-conditioning engineers. Both need to have available a range of laboratory methods for recognizing the infections and discovering the microbes. Dr Harrison and Dr Taylor have gathered a team with practical experience in a wide variety of microbiological methods now available. This volume certainly provides very valuable information and instruction on the methods that are already applicable in diagnostic laboratories, and at the same time indicates the scope of other techniques that are presently confined to research laboratories but will undoubtedly become more widely applicable in the future.

R. E. O. WILLIAMS

Plush, Dorset
October, 1987

Preface

The published literature on the subject of *Legionella* and Legionnaires' disease is now extensive and in addition to numerous contributions in a variety of scientific journals this includes several substantial books. We feel however that there is a need for a laboratory handbook for microbiologists which presents the available technical information concerning the demonstration, isolation, culture and identification of *Legionella* together with methods for the serological diagnosis of legionellosis in patients.

This book is based on the practical experience of the various contributors and recommends procedures that have been well tried in their laboratories. The intention has been to give adequate practical information so that the methods described can be followed with minimal referral to the widely dispersed original scientific papers.

The experience on which the various methods have been evaluated could only have been gained with the help and enthusiasm of a myriad of colleagues including epidemiologists, clinicians, nurses, medical and non-medical microbiologists, medical laboratory scientific officers, hospital and other engineers, water authority personnel, librarians and others. We gratefully acknowledge their help and also that of North American colleagues especially in the Centers for Disease Control, Atlanta.

The book contains photographic illustrations for which we are indebted to Mr John Gibson, Chief Medical Photographer here at the Central Public Health Laboratory and drawings including GLC profiles prepared by Penny Forsyth. The secretarial assistance of Mrs Corrina Scott is also gratefully acknowledged.

T. G. HARRISON
A. G. TAYLOR

Colindale, September 1987

A Laboratory Manual for *Legionella*
Edited by T. G. Harrison and A. G. Taylor

CHAPTER 1

Introduction

A. G. Taylor

More than ten years after the discovery of *Legionella pneumophila* the diagnosis of Legionnaires' disease is still established in the majority of cases by serological methods. This situation is unusual among bacterial infections which are usually diagnosed by culture of the causative organism from appropriate clinical specimens. Where this is not the case alternative methods are used because of difficulties associated with the *in vitro* culture of the organism, for example *Treponema pallidum* and *Mycoplasma pneumoniae*. In the case of *L.pneumophila*, difficulties were at first encountered with culture methods and isolation was often only achieved using guinea-pigs. However, another problem has also contributed to the dependence on serological methods, viz. the possible acquisition of infection by laboratory workers through the handling of clinical specimens.

The reluctance to introduce culture methods for *Legionella* in routine diagnostic bacteriology laboratories in the UK may to some extent have been due to the Godber Report (DHSS, 1975). This report was followed in 1978 by the 'Howie Code' (DHSS, 1978) produced by a Working Party under the chairmanship of Sir James Howie. In this publication 'the causal organism of Legionnaires' disease' was included in Category B1. This category included organisms considered to offer 'special hazards to laboratory workers and for which special accommodation and conditions for containment must be provided'. Hence the isolation of *L.pneumophila* was not attempted in many routine diagnostic laboratories. In addition, the requirement for special culture media and the popular image of the organism as a dangerous pathogen have probably contributed to the reliance on methods other than culture.

At the present time microbiologists are well aware of the lack of evidence for laboratory infections caused by *Legionella*, and of the absence of person-to-person transmission of Legionnaires' disease. Special containment facilities are no longer required, nevertheless in the UK approximately 90% of cases of Legionnaires' disease are still diagnosed without culture of the causative organism. One of the principal aims of this book is to indicate simple methods for the culture of *Legionella* (Chapter 3) which are known to be successful and to encourage microbiologists in routine diagnostic laboratories to use these in addition to serological and microscopic methods.

The methods for the isolation of *Legionella* from environmental specimens are not identical to those appropriate for culture of patient specimens. Different selective media and specimen pretreatments are required and these, together with advice on obtaining appropriate samples, are described in Chapter 4. The provisional identification of isolates from any source as belonging to the genus *Legionella* is possible using simple phenotypic characteristics (Chapter 5). Further phenotypic characteristics and serological methods (Chapter 6) can then be employed to permit the species to be determined. Confirmation of the identity of the strain by the study of fatty acid and isoprenoid quinone profiles is recommended (Chapter 7), although many laboratories may not have the resources to implement these specialized techniques.

Demonstration of legionellae in clinical specimens by immunofluorescence microscopy or of antigens by immunoassay (Chapter 8) is of value in that an extremely rapid diagnosis can often be made which may be critical in patient management. These techniques are also important in the examination of fixed clinical material in retrospective diagnosis and can be helpful in the selection of clinical or environmental specimens for isolation and culture of legionellae.

Prevention of many cases of Legionnaires' disease is now possible because of our current understanding of its epidemiology. The investigation of the Philadelphia outbreak in 1976 showed the causative organism to be airborne and its presence in air conditioning systems was demonstrated retrospectively on stored specimens from the 1968 outbreak of Pontiac fever (Glick *et al.*, 1978) and during investigations of various outbreaks of the typical form of Legionnaires' disease in the USA (Broome & Fraser, 1979). The organism's presence and significance in plumbing systems has been well documented since Tobin and colleagues (1980) in Oxford, England isolated *L.pneumophila* serogroup 6 from both a patient and the hospital shower she used. The isolation of a strain of *L.pneumophila* of an uncommon serogroup from both patient and the environmental source was helpful in establishing this route of infection. Today more sophisticated typing methods are available for comparing clinical and environmental strains thus enabling sources of infection to be identified and these are discussed in Chapter 10.

It is possible, with our present knowledge of the control of legionellae in water and cooling systems, that outbreaks of Legionnaires' disease may become rare events. However, the number of travel-associated and nosocomially acquired sporadic cases is likely to increase, thus the need for laboratory diagnosis and environmental surveillance will remain.

The current developments in nucleic acid probe technology, in addition to the development of typing methods unaffected by cultural conditions, will undoubtedly have an impact on diagnostic procedures and in the identification of species. While it is unlikely that these methods will entirely replace bacterial culture and isolation, the eventual development of sensitive non-radioactive labels will ensure their use in the routine diagnostic laboratory. It is hoped that meanwhile this book, which brings together the practical experience of a number of workers in the field, will prove useful in the laboratory.

A Laboratory Manual for *Legionella*
Edited by T. G. Harrison and A. G. Taylor

CHAPTER 2

Clinical Features of Legionnaires' Disease

C. Mayaud and E. Dournon

I Introduction

Legionnaires' disease (LD) is essentially an acute infection of the lungs caused by Gram-negative bacilli belonging to the genus *Legionella*. The disease was clinically and bacteriologically first characterized following the Philadelphia outbreak in July 1976 (Fraser *et al.*, 1977; McDade *et al.*, 1977). However, it is clear that LD is not a 'new disease' since cases of pneumonia which occurred during the 1940s have been shown retrospectively to be due to *Legionella*. The first recognized cases of LD were associated with a large outbreak and high mortality, hence it is generally thought to be a very severe form of pneumonia. In fact it should be strongly emphasized that when diagnosed early and appropriately treated LD is often easily cured. Fortunately, in most instances the clinical picture of LD allows sufficient suspicion of the diagnosis to initiate appropriate therapy immediately after taking specimens for laboratory confirmation.

LD is a rare cause of pneumonia and in most areas probably accounts for less than 5% of all pneumonia cases requiring admission to hospital (Woodhead *et al.*, 1987). Much higher figures have been reported but in these series diagnosis relied only on serology using antigens whose specificity has not been established or is known to be poor (Wilkinson *et al.*, 1983; Fallon & Abraham, 1983; Macfarlane *et al.*, 1982). In certain patient groups, such as organ transplant patients, the incidence is much higher and LD may be the main cause of lung infections (Dournon *et al.*, 1983; Meyer, 1984; Kirby *et al.*, 1980; Dowling *et al.*, 1984b).

The disease is most often seen in particular groups of adults. However, it is probable that any person can be infected with *Legionella* including, on rare occasions, children. In our series of patients (644 cases from hospitals throughout the Paris area), 72.2% were male (mean age 52.6 yrs, median 54 yrs, range 0.3–90 yrs), and 27.8% female (mean age 56.1 yrs, median 58 yrs, range 0.1–95 yrs), a male to female ratio of 3:1. The immune status of 526 of the patients was known precisely: 219/526 (41.6%) were immunocompromised (Table 2.1), and 307 (58.4%) were not. Nearly all of the non-immunocompromised patients (76.5% males, mean age 55.2 yrs, 23.5% female, mean age 63.5 yrs) were associated with at least one of the following risk factors: age more than 70 years, recent surgery, congestive heart failure, chronic bronchitis, liver cirrhosis, renal insufficiency, or heavy smoking.

Travelling may also be a risk factor, especially for British tourists visiting southern European countries. Particular occupations have not been clearly demonstrated to be risk factors. However, workers exposed to air cooling tower aerosols should probably be considered at increased risk.

Epidemics of LD account for only a minority of cases and these can often be linked to a common source, for example exposure to aerosols containing the bacteria produced by an air cooling tower, or exposure to aerosols of contaminated hospital, hotel or other domestic water. Sources of sporadic cases are more difficult to recognize than those of outbreak-associated cases and probably include private residences. It should be stressed that in some instances it is likely that apparent sporadic cases may be linked to a common but unrecognized environmental source. In the UK and United States, sporadic cases of LD have been reported to occur mainly from spring to late autumn, however in northern France they are seen throughout the year.

Nosocomially acquired LD may occur as epidemic or sporadic cases. The latter can be demonstrated in most hospitals if appropriate investigations of patients with nosocomial pneumonia are made. The maximum incubation period for LD is estimated to be 12 days, therefore any such infection arising in a patient admitted to hospital for an unrelated reason 12 or more days earlier is considered to be nosocomially acquired. Cases appearing less than 12 days after admission should be considered as possibly nosocomial. In our series of 644 patients with LD, 196 (30.4%) were in patients admitted to hospital at least 10 days before onset of LD.

Table 2.1 Underlying diseases in 219 immunocompromised patients with Legionnaires' disease in Paris.

MALIGNANCIES		99
Leukaemia	51	
– chronic lymphocytic leukaemia : 16		
– hairy cell leukaemia : 13		
– acute myeloblastic leukaemia : 10		
– acute lymphoblastic leukaemia : 8		
– chronic myeloid leukaemia : 4		
Cancers	27	
– lung : 8		
– other : 19		
Lymphoma	21	
IMMUNOSUPPRESSIVE THERAPY		110
Transplant patients	56	
*Other**	54	
IMMUNE DEFICIENCY		10
Primary	7	
AIDS	3	

* Patients with asthma, multiple sclerosis, Goodpasture's syndrome, systemic lupus erythematosus, treated with steroids alone or in combination with cyclophosphamide or azathioprine.

Airborne infection is the only documented route of infection but this does not exclude other possibilities. The currently accepted sources of the bacterium are tap water and air conditioning systems, especially air cooling towers. No person-to-person transmission has been reported.

Although there are at least 28 *Legionella* species, some of which have more than one serogroup, it should be remembered that not all of these are necessarily present throughout the world and that not all legionellae are pathogenic for man. Many species have not been isolated in Europe although extensive systematic environmental studies have been performed in both the UK and France. Moreover, some species found in the USA have only been isolated from environmental samples. Where proven by culture, infections caused by any *Legionella* other than *L.pneumophila* serogroup 1 have almost always been seen in patients who were immunocompromised, usually severely. In Paris we have cultured 3700 clinical samples and isolated 157 strains of *Legionella*, all of which were *L.pneumophila*. Of these 144 were *L.pneumophila* serogroup 1 (43% patients were immunocompromised), and the remaining 13 were *L.pneumophila* serogroups 3, 4/5, or 6 (all patients were immunocompromised).

Pontiac fever (Glick *et al.*, 1978) is a self-limiting influenza-like non-pneumonic form of the disease which has been related to the inhalation of *Legionella* on epidemiological and serological data. The incubation period is short (usually 36–48 hours) and the attack rate high (95% in the Pontiac outbreak). The illness resolves spontaneously in 2–5 days. The chest radiograph is negative and laboratory findings normal although a seroconversion to *Legionella* may be demonstrable.

II Clinical features

A Onset

The onset of LD is generally less abrupt than in typical pneumococcal pneumonia. Fever, headaches, weakness and malaise are the first manifestations. In about half of the cases gastrointestinal abnormalities, including nausea, vomiting, abdominal pain and/or diarrhoea, are present. There is no sign of upper respiratory tract involvement and at this early stage there is often no obvious sign of lung involvement (Meyer, 1984; Kirby *et al.*, 1980; Mayaud *et al.*, 1984).

B Acute phase

i Clinical signs and symptoms

The patient's body temperature is high, greater than 40 °C in many cases, and the pyrexia unremitting in nature. Rigors and relative bradycardia are present in about 75% and 60% cases, respectively.

Respiratory symptoms often only appear later in the course of the disease. A cough is rare and if present it is non-productive early in the illness but later small amounts of purulent, and sometimes bloody sputum, may be expectorated. Lung examination usually reveals abnormalities at this stage of the disease with at least rales and very often lung consolidation being evident. A small pleural effusion, sometimes suspected from the presence of thoracic pains, is present in about one-third of patients but is difficult to recognize on clinical examination. Progression of the lung infection may be rapid and be responsible for increasing dyspnoea and acute respiratory failure.

A peculiar, although non-specific, feature of LD is the presence in many cases of gastrointestinal symptoms or neurological signs. Diarrhoea, present in about 50% cases, appears early in the course of the disease and if watery is especially suggestive of LD. Nausea, vomiting, or right lower abdominal pain with tenderness are each noted in 20–30% patients. Nervous system manifestations are dominated by alterations of the mental status: disorientation, confusion, agitation or obtundation are present in about 25% cases. Such cases can easily be misdiagnosed as pneumococcal pneumonia together with delirium tremens. These neurological manifestations are of considerable diagnostic value when they cannot be explained by hypoxaemia or by meningitis. Typically the CSF is normal in patients with LD. Many other neurological abnormalities have been reported but these are very rare.

ii Radiological findings

Lung opacities in LD always have an alveolar-filling pattern (Figure 2.1). Lobar, patchy, or subsegmental infiltrates or dense consolidations are the most frequent radiographic features. The pneumonia is initially limited to one or several foci, especially in immunocompromised patients, but has a marked tendency to extend to adjacent and non-adjacent areas. In about 60% of cases the pneumonia is bilateral. Extension of the lung opacities during the first days of treatment is common while cavitation is possible especially in immunocompromised patients treated late.

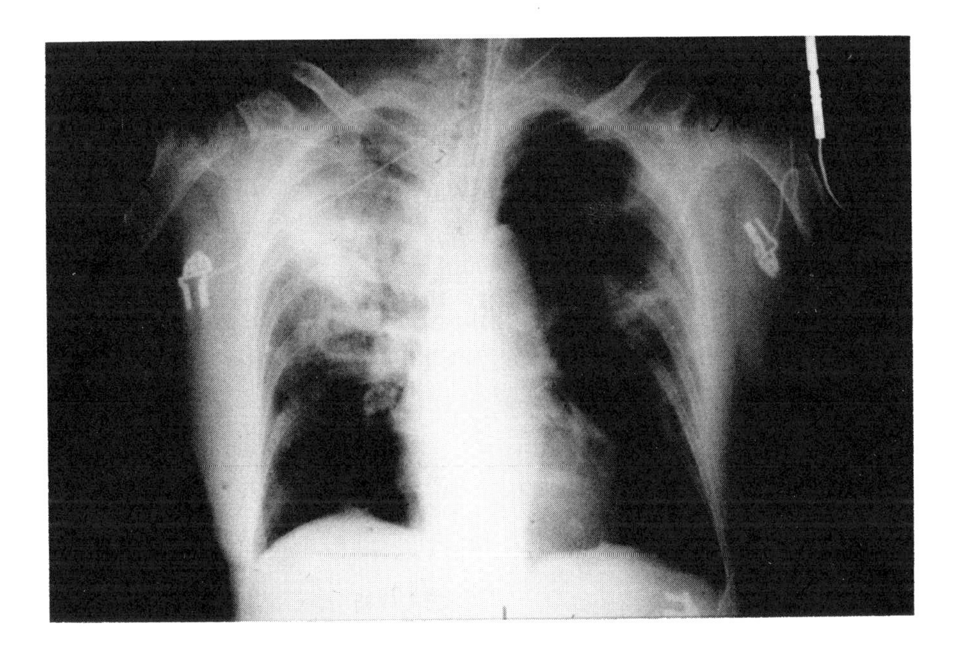

Figure 2.1

iii Routine laboratory findings

The white blood cell count is not helpful since hyperleucocytosis, normal count or leucopenia can be observed. The erythrocyte sedimentation rate is almost always >100 mm/h. Hyponatraemia and hypophosphoraemia are not unusual but are seen in other forms of pneumonia. Liver abnormalities are frequently found with elevated serum levels of transaminases and/or alkaline phosphatase or, in some cases, of bilirubin. Proteinuria, haematuria and renal insufficiency are infrequent in patients receiving early treatment. Elevation of creatinine phosphokinase and aldolase blood levels, suggestive of muscle damage, is seen in some patients.

iv Diagnosis

None of the above mentioned signs and symptoms is specific of LD. However, LD can be strongly suspected if one takes into account the following factors:

- age, sex, immune status of the patients;
- presence of clinical or biological extra-thoracic abnormalities;
- chest X-ray findings;
- negativity of conventional bacteriological investigations;
- resistance to treatment with all beta-lactam and aminoglycoside antibiotics.

Other diseases that may closely mimic LD are the severe forms of pneumonia caused by *Mycoplasma pneumoniae* or *Chlamydia psittaci*. Fortunately erythromycin, the reference drug for LD, is also active against both these bacterial agents.

III Outcome

A Treated patients

In our experience patients treated with appropriate antibiotics within seven days of onset will recover in most instances even if they are immunocompromised (Mayaud *et al.*, 1984). Fever usually abates after 36–72 hours of treatment but the return to normal blood gas levels is often delayed for a few days and chest X-ray abnormalities may only resolve after several weeks.

B Untreated patients

The prognosis is poor in patients not treated before acute respiratory failure and shock develop. This is usually late in the course of the illness. In such cases rhabdomyolysis, acute renal failure, pancytopenia, disseminated intravascular coagulation, icterus or coma are not infrequent. However, in most instances patients die from the respiratory distress syndrome and not from the extra-pulmonary manifestations of the disease.

IV Antibiotic treatment

The conventional *in vitro* means of appraisal of antibiotic activity, such as minimum inhibitory concentration or minimum bactericidal concentration, are not predictive of the *in vivo* efficacy of antibiotics given in LD for the following two reasons:

- *Legionella* are facultative intracellular parasites (Horwitz & Silverstein, 1980), and thus some antibiotics that are active *in vitro* are not active *in vivo* because their cellular permeability is poor, e.g. beta-lactam antibiotics (Vildé *et al.*, 1986; Horwitz & Silverstein, 1983).
- the media used to culture *Legionella* inhibit the activity of some antibiotics (Edelstein & Meyer, 1980; Dowling *et al.*, 1984a).

Because of these two obstacles, and also the relative rarity of LD and the lack of prospective studies, the optimal antibiotic treatment of LD has not been established. However, a limited number of antibiotics have been shown to be active in animal and cell models of infection (Vildé *et al.*, 1986; Edelstein *et al.*, 1984b; Dournon *et al.*, 1986c) and in patients. These include two macrolides (erythromycin and spiramycin), rifampicin and a new fluoro quinolone, pefloxacin. In an *in vitro* model of *Legionella* infection any combination of these drugs has a synergistic effect (Vildé *et al.*, 1986). Although there are no controlled clinical data to substantiate this observation all experienced clinicians recommend such a regimen in severe cases.

V Conclusions

The clinical diagnosis of LD is not difficult when this possibility is kept in mind. Early treatment, including the administration of erythromycin, should be initiated as soon as reasonable clinical suspicion of such a diagnosis exists. A patient with LD, even if severely immunocompromised, has an excellent prognosis when treated promptly.

In addition, the routinely available techniques for laboratory confirmation allow this diagnosis to be confirmed within a few hours, by positive immunofluorescence on clinical specimens, or a few days, by culture and serology.

A Laboratory Manual for *Legionella*
Edited by T. G. Harrison and A. G. Taylor

CHAPTER 3

Isolation of Legionellae from Clinical Specimens

E. Dournon

I Introduction

Paradoxically, although Legionnaires' disease (LD) was first recognized only ten years ago, laboratory diagnosis of this disease is now easier than the diagnosis of many other lung infections. The specificity and sensitivity of some serodiagnostic tests for *L.pneumophila* serogroup 1 infections are without parallel among bacterial pneumonias and these together with the demonstration of *Legionella* in clinical specimens by culture and immunofluorescence (IF) allow a diagnosis to be established rapidly in almost all cases.

There are many reasons for attempting to culture *Legionella* from clinical specimens:

- Isolation of *Legionella* is proof of the diagnosis, whereas antibody detection and demonstration of the bacterium by IF are open to doubt regardless of the high specificity attainable with these tests.
- Isolation of the bacterium may in some cases be the only means to establish a diagnosis of LD, e.g. in patients who do not produce antibodies and those infected with a non-*L.pneumophila* serogroup 1 strain. In the latter serology and IF have not been adequately evaluated and therefore cannot be used as reliable diagnostic tools. Thus culture is the only means to recognize all *Legionella* spp., including as yet unknown species and serogroups.
- Isolation of *Legionella* sometimes allows early diagnosis when serology may still be inconclusive.
- Subgrouping of the strains, especially with monoclonal antibodies, allows epidemiological studies to be undertaken. These are of considerable interest in the recognition of environmental sources of *Legionella* associated with disease.
- Collections of strains are also necessary for the evaluation of other typing systems.
- A large number of culture-proven cases from whom serum and other specimens may be obtained is necessary to evaluate both the sensitivity and specificity of any future diagnostic methods.
- Repeated isolation attempts in patients receiving treatment allows the efficacy of the treatment to be monitored. This may help to define better antibiotic regimens for the treatment of LD.

In the past various reasons have been given for not attempting the culture of

Legionella. However, during the last ten years considerable progress has been achieved in this field and these reasons are no longer valid. For example:

- No special laboratory facilities or equipment are required.
- Culture media can be easily prepared in any laboratory and are now commercially available.
- The routine identification of isolated strains is usually rapid and uncomplicated.
- Isolation of *Legionella* from any respiratory tract or systemic specimen is diagnostic of LD as colonization without infection has never been described.
- The proportion of positive cultures obtained when appropriate specimens are available from cases of LD compares favourably with that obtained from other pneumonias, including those caused by *Streptococcus pneumoniae*.

In summary, culture diagnosis of LD is at least as useful as culture diagnosis of many other diseases and technically can be achieved in any routine diagnostic bacteriology laboratory.

The procedures for the culture of *Legionella* from clinical specimens presented in this chapter are based both on the published literature and on our own experience. Our intention is to give practical information, and not to review in detail all methods that have been described.

II Patient selection

Legionnaires' disease is not a very common form of pneumonia and probably comprises less than 5% of acute pneumonia cases requiring hospital admission. Careful analysis of patient characteristics (age, sex, immune status), clinical features, chest radiology and the effect of antibiotics already administered, can often indicate or exclude a preliminary clinical diagnosis of LD (see Chapter 2). For these reasons it is not necessary that laboratories accept requests to culture *Legionella* systematically on all patients with 'chest infection'.

III Specimens suitable for culture

As LD is primarily a pneumonia, specimens originating from the lower respiratory tract are the most appropriate. However, severe forms of the disease are often associated with bacteraemia, and *Legionella* can sometimes be demonstrated in various sites, especially from specimens taken post-mortem.

A Lower respiratory tract (LRT) specimens

Any such specimens, including expectorated sputum, are acceptable. The rate of culture positivity in our laboratory, according to the type of LRT specimen examined is summarized in Table 3.1. It should be noted that these data do not allow an accurate comparison of the different samples since they do not originate from similar patient groups. Endotracheal aspirates are generally obtained from severely ill patients undergoing mechanical ventilation, whereas expectorated sputum and transtracheal aspirates are usually obtained from less severe cases. Furthermore, performing bronchoalveolar lavage (BAL) or transtracheal aspiration (TTA) depends on the clinical situation and on the physician's habits.

Several points should be noted:

1. Pleural fluids rarely yield positive results.
2. Transtracheal aspirates and bronchial brushing specimens are not ideal since the volume of material available for culture and IF is often very limited. However, this disadvantage is often outweighed by the absence of contaminants which allows *Legionella* to be isolated even when present in only very small numbers.
3. If sputum expectorated early in the course of a pneumonia is purulent a diagnosis of LD is very unlikely. However, later in the illness sputum may also be purulent in LD patients.
4. Saline, which is generally used for TTA and BAL, is deleterious to *Legionella*.
5. Lidocaine, which is often used for local anaesthesia during lower respiratory tract sampling procedures, has been shown to be inhibitory for many other microorganisms (Winberley *et al.*, 1979) and may inhibit *Legionella*.
6. In our experience sputum expectorated just after taking a TTA is an excellent specimen to culture for *Legionella*.

Table 3.1 The proportion of first specimens taken from 217 patients with LD which were culture positive or immunofluorescence (IF) positive.

	Number	Specimens positive by:		
		Culture	IF	Culture or IF
Endotracheal aspirates[1]	98	68 (69.4%)	50 (51.1%)	75 (76.5%)
TTA[2]	16	9 (56.2%)	7 (43.7%)	11 (68.7%)
BAL[3]	36	18 (50.0%)	14 (38.5%)	24 (66.7%)
Sputum	67	22 (32.8%)	21 (31.3%)	37 (55.2%)
Total[4]	217	117 (53.9%)	92 (42.4%)	147 (67.7%)

1. Any bronchial or tracheal aspirate.
2. Transtracheal aspirate (TTA).
3. Bronchoalveolar lavage (BAL).
4. Definition of a case: an illness compatible with LD together with a positive culture, and/or positive serology as defined in Chapter 9, and/or IF positive on specimens other than sputa.

B Blood

In severe forms of LD, especially in immunocompromised patients, bacteraemia has been shown to occur. In one published series this was demonstrated in 6/16 (37.5%) patients (Rihs *et al.*, 1985), and in our own series in 23/46 (50%) patients (Dournon *et al.*, 1986b). These figures are most probably underestimations of the actual frequency of bacteraemia in such severely ill patients, since in many cases only a single specimen could be tested. Furthermore, many of the patients had already received antibiotics known to be active on *Legionella* before blood was taken.

Culture from blood should therefore be attempted in patients clinically likely to be infected with *Legionella* since:

1. Blood collection is an innocuous procedure which can easily be repeated in order to increase the possibility of successfully isolating *Legionella*.
2. In patients with demonstrated bacteraemia, repeated blood cultures allow the success of the antibiotic treatment to be assessed.

It is noteworthy that retrospective diagnosis using seeded conventional blood culture bottles has been demonstrated (Farrington & French, 1983).

C Other specimens

Other extra-pulmonary sites from which *Legionella* has been cultured are:

Liver and spleen	(these are commonly positive in patients dying from LD).
Heart (endocarditis)	(McCabe *et al.*, 1984)
Pericardial fluid	(Mayock *et al.*, 1983)
Kidney	(Dorman *et al.*, 1980)
Peritoneal exudate	(Dournon *et al.*, 1982)
Gut (terminal ileum)	(Dournon *et al.*, 1982)
Perirectal abscess	(Arnow *et al.*, 1983)
Muscle	(Dournon, unpublished data)
Wounds	(Brabender *et al.*, 1983)
Cutaneous abscess	(Ampel *et al.*, 1985)
Haemodialysis shunt	(Kalweit *et al.*, 1982)

In addition *Legionella* has been demonstrated by IF only, in the brain (Cutz *et al.*, 1982), faeces (Dournon, unpublished data) and a biopsy taken from a maxillary sinus (Schlanger *et al.*, 1984). To our knowledge *Legionella* has not been isolated from CSF, however the number of culture attempts is probably too small to exclude the possibility that the organism could sometimes be present. Most attempts to isolate *Legionella* from non-respiratory tract specimens have been made using post-mortem specimens rather than during life. Therefore, the possibility of a *Legionella* infection should be considered in patients with endocarditis and negative blood cultures, or in any local suppuration involving Gram-negative bacilli which are found not to grow on conventional media. In such cases specimens should be collected and attempts made to isolate the organism.

IV Optimal timing of specimen collection

In our series of 247 patients with LD, in whom at least one attempt to culture *Legionella* was made, the organism was isolated from the first specimen in 124, and from a subsequent specimen in 11 additional patients (four endobronchial samples, six blood cultures and one post-mortem lung). Thus when culture of the first specimen is negative subsequent specimens are only occasionally positive. These results are in agreement with previously reported data (Edelstein, 1984b). Nevertheless, in patients clinically suspected of having LD, especially if a first specimen is IF negative, blood and a second LRT specimen should be obtained for culture. This is especially true in the case of immunocompromised patients investigated early in the course of the disease when the foci of infection are limited to one or several small nodules in the lung. In this situation, IF and culture are likely to prove negative initially but may be positive 12–48 hours later when the pneumonia has extended. This appears to be true irrespective of whether or not appropriate treatment has been initiated.

Ideally specimens for culture should be taken before antibiotic treatment is initiated. However, it should be strongly emphasized that *Legionella* is very commonly isolated from LRT specimens, or even from the blood, of patients who have been treated for several days with erythromycin alone or in combination with rifampicin. This is clearly illustrated by the data presented in Table 3.2.

In severe cases, for example patients requiring mechanical ventilation, daily examination for *Legionella* in LRT and blood specimens should be undertaken. This may permit early recognition of treatment failure and can indicate the need to review the dosage and nature of chemotherapy. In immunocompromised patients with lung cavitation due to *Legionella*, the persistence of the organism is not unusual despite clinical improvement of the patient.

V Transport of specimens

Legionellae are not delicate pathogens and do not require any special transport procedures. *L.pneumophila* has been re-cultured from clinical specimens kept at 4 °C for periods from several weeks to over 18 months. However, specimens should not be allowed to dry out and if necessary a small volume of sterile and filtered distilled water can be added to prevent this. Other solutions should be avoided, especially saline which is inhibitory for *Legionella*.

Table 3.2 Number and percentage of culture-positive specimens according to the number of days of appropriate active treatment (i.e. erythromycin, rifampicin, or pefloxacin given alone or in combination).

Days of treatment	Endotracheal aspirates +/N[3] (%)	TTA[1] BAL +/N(%)	Sputum +/N(%)	PM[2] Lung +/N	Other +/N	Total +/N(%)
3	70/94 (75)	17/32 (53)	24/47 (51)	6/7	4/9	121/189 (64)
3 – 7	24/51 (47)	2/6	6 – 111	6 – 10	7/8	45/86 (52)
8 – 12	3/26 (12)	ND[4]	1/8	1/2	0/5	5/41 (12)
13 – 18	4/18 (22)	0/2	1/4	1/2	0/2	6/28 (21)
19 – 27	3/23 (13)	ND	0/4	1/2	ND	4/29 (14)
28 – 50	0/8	0/2	0/7	0/2	0/3	0/22

1. TTA = transtracheal aspirate, BAL = bronchoalveolar lavage.
2. PM Lung = post-mortem lung tissue.
3. +/N = Number of positive cultures/number of specimens cultured.
4. ND = No data.

The most important aim is to prevent overgrowth of specimens by bacteria or yeasts which are often inhibitory to *Legionella*. Specimens that cannot be cultured immediately are best refrigerated (4 °C) and if a delay of more than 3–4 days is expected freezing (–20 °C or lower) is recommended. However, *Legionella* has often been isolated from specimens kept at room temperature for several days and such specimens including those received through postal services should not be rejected.

VI Processing of specimens

A Selective procedures

Several procedures have been described to increase the likelihood of isolating *Legionella* from clinical samples:

i Heat treatment

Legionella is generally more tolerant to heating than are many other bacteria frequently found in clinical specimens. Heating is especially useful when the specimens contain large numbers of yeasts or bacteria such as *Pseudomonas* spp. and *Proteus* spp. which may not be inhibited by the antibiotics included in the selective media. Heating specimens in a water bath at 60 °C for one to three minutes or at 50 °C for 30 minutes significantly reduces the numbers of these bacteria without greatly reducing the numbers of *Legionella* (Edelstein *et al.*, 1982b; Dennis *et al.*, 1984a). Unfortunately some strains of *Legionella* species may also be inhibited by these procedures.

ii Acid treatment

Legionella are more resistant to low-pH exposure than are many other bacteria. Thus, low-pH treatment, which is especially useful for recovery of *Legionella* from environmental samples, may also be of value when clinical specimens are heavily contaminated (Greaves, 1980; Bopp *et al.*, 1981; Buesching *et al.*, 1983).

Reagents

Buffer pH 2.0
0.2 M KCl: 14.91 g/l in distilled water
0.2 M HCl: 17.2 ml/l (36% HCl) in distilled water
18 parts of 0.2 M KCl are added to one part of 0.2 M HCl
After checking the pH the buffer is dispensed into screw-cap bottles (1–3 ml) and sterilized by autoclaving.

Neutralizing solution
0.1 N KOH: 6.46 g/l in distilled water
10.7 ml is diluted with 89.3 ml distilled water, dispensed into screw-cap bottles (1–3 ml) and sterilized by autoclaving.

Procedure

The sample is placed in a sterile tube and an equal volume of buffer pH 2.0 added. The mixture is vigorously vortexed and left to stand for 5–20 minutes (dependent on the level of contamination). A volume of the neutralizing solution equivalent to the original sample volume is added and vortexed. The acid-treated specimen can then be inoculated directly on to a culture plate.

An alternative acid-treatment procedure, which omits the neutralizing solution, has been successfully used and is described in Chapter 4.

iii Dilution of specimens

The dilution of specimens reduces the concentrations of contaminants, of natural inhibitors of *Legionella* and of antibiotics when present. Dilutions ranging from 10^{-1} to 10^{-4} should be made in distilled water or in Mueller–Hinton broth.

iv Selective media

The BCYE medium can be rendered partially selective by the addition of antibiotics (BCYEA) that are not strongly inhibitory to the growth of *Legionella*. BCYEA agar plates are especially useful for heavily contaminated specimens such as expectorated sputa.

Several antibiotic combinations have been described. The one we describe below inhibits the growth of many bacteria commonly found in clinical specimens without greatly reducing the yield of *Legionella*. Some authors omit vancomycin which has been found in one study to be quite inhibitory to *Legionella* (Edelstein, 1981). This is not our experience (Table 3.3), and we consider vancomycin to be useful because it inhibits those staphylococci which are not inhibited by cefamandole.

B Preparation of specimens

Bronchoalveolar lavages and pleural fluids should be centrifuged (>200*g* for 30 minutes) and the pellet inoculated on to the appropriate media. For other respiratory tract specimens a milky or bloody part should be picked up with a sterile Pasteur pipette and used as the inoculum.

Tissues, e.g. lung biopsies and post-mortem lung, should be ground with sterile sand and water using a pestle and mortar (1–5 g of tissue in 2–5 ml distilled water). The mixture is then transferred to a sterile tube, vortexed and left for a few minutes to let the sand sediment. The tissue homogenate is then plated as described below.

C Plate inoculation and incubation

After making appropriate dilutions in distilled water, culture plates are inoculated with 0.1 ml of the neat or diluted specimen. In practice the inoculum size can be slightly increased or decreased according to its known or expected degree of contamination. Culture plates should be incubated at 35–37 °C, in humidified air or in air plus 2.5% CO_2.

D Other practical considerations

The degree of contamination of clinical specimens by bacteria and yeasts is critical since *Legionella* growth is easily inhibited by other microorganisms. Contamination has to be evaluated as precisely as possible to decide whether or not to use one or several of the selective procedures described above. In order to do this it is often useful to consider the following points:

1. The nature of the specimen: TTA, BAL, tissue biopsies, pleural fluid and blood are generally not contaminated.
2. The delay and storage conditions between the sample being taken and its being processed in the laboratory.
3. The number of days a chest tube has been *in situ*: after 5–7 days the likelihood of respiratory tract colonization with *Pseudomonas* spp., Enterobacteriaceae and/or *Candida* spp. is high.
4. The antibiotics used to treat the patient: (a) broad spectrum antibiotics which are not active against *Legionella* (beta-lactams and aminoglycosides) given for a few days may help the recovery of *Legionella* by reducing the numbers of other bacteria which would otherwise be present in the sputum; (b) the administration for a few days of antibiotics which are active against *Legionella* does not generally prevent the recovery of *Legionella*. However, *Legionella* that have been exposed to such antibiotics often fail to grow on BCYEA and are more susceptible to heat or acid treatments. Isolation is therefore best attempted by specimen dilution.
5. Gram staining of the specimens may give a good indication of the level of contamination.
6. The numbers of *Legionella* estimated to be present by IF. For example if 20 organisms are seen in a high power field the sample can be diluted to 10^{-4}, conversely if the IF was negative it would be wise to dilute the specimen with caution.

From our own experience, and as reported in the literature, specimen dilution and selective media should be used routinely especially for heavily contaminated samples such as sputum (Edelstein, 1984b). Conversely, heat and/or acid treatments are indicated in only a limited number of cases, for example patients with an endotracheal tube heavily colonized with Gram-negative bacteria. These treatments can prevent successful culture of *Legionella* in a significant number of instances and therefore must only be used in addition to direct plating.

Until recently we plated all specimens neat and diluted to 10^{-2} and 10^{-4} on both BCYE and BCYEA (six plates per specimen). Our results, according to the kind of specimen, dilution and medium used, are summarized in Tables 3.3. and 3.4 from which it can be concluded that:

- If possible, a specimen should be inoculated both neat and diluted to 10^{-2} on BCYE and BCYEA (four plates per specimen).
- Inoculation of specimens diluted to 10^{-4} is not routinely indicated. However, if BCYEA plates are not available inoculation of the specimen neat and diluted to 10^{-2} and 10^{-4} on BCYE medium is recommended.
- If the number of plates used for each specimen is limited by cost then contaminated specimens should be inoculated diluted to 10^{-2} on BCYE and neat on BCYEA whereas 'clean' specimens can be inoculated neat on to both culture media.

Table 3.3 Comparison of media performance in the isolation of *Legionella* from respiratory tract specimens.

	No difference between BCYE and BCYEA[a]	BCYE superior to BCYEA[b]	BCYEA superior to BCYE[c]
Endotracheal aspirates $n = 90$	45.6	23.3	31.1
TTA or BAL $n = 23$	60.9	17.4	21.7
Expectorated sputum $n = 25$	24.0	4.0	72.0

a. Percentage of specimens culture positive on both BCYE and BCYEA plates. Number of colonies within one log of each other.
b. Percentage of specimens culture positive only on BCYE plates, or number of colonies on BCYE at least 1 log greater than on BCYEA plates.
c. Percentage of specimens culture positive only on BCYEA plates, or number of colonies on BCYEA at least 1 log greater than on BCYE plates.

Table 3.4 The effect of specimen dilution on the isolation rate.*

	Neat	Diluted 10^{-2}	Diluted 10^{-4}	No difference
Endotracheal aspirates $n = 112$	55.3	28.6	2.7	13.4
TTA or BAL $n = 25$	68.0	8.0	0.0	24.0
Expectorated sputum $n = 38$	39.5	39.5	5.2	15.8

* Expressed as the percentage of specimens for which a particular dilution was optimal. A dilution was considered optimal whenever:

- it was the only dilution from which *Legionella* was cultured, or
- the number of colonies obtained was 10 times greater than that obtained from other dilutions, or
- the colonies were visible more than 24 hours earlier than on the other plates.

VII Demonstration of *Legionella* in blood

Historically the first clinical isolation of *Legionella* was made from a specimen of blood by Jackson in 1947 (McDade *et al.*, 1979). Since the Philadelphia outbreak there have been only a few reports of isolation from blood (Macrae *et al.*, 1979; Edelstein *et al.*, 1979; Meyer *et al.*, 1980; Farrington & French, 1983; Chester *et al.*, 1983). However, in two recent prospective studies bacteraemia was shown to occur in a significant number of cases (Rihs *et al.*, 1985; Dournon *et al.*, 1986b). Several techniques have been used to isolate *Legionella* from blood:

1. Direct seeding of liquid or biphasic media.
2. Seeding of BACTEC system blood culture and subculture on BCYE plates (Rihs *et al.*, 1985).
3. Seeding of BCYE plates after lysing blood cells.

It is not known which of the above methods is to be preferred as no comparisons have been reported. We describe here the third method as no special material or equipment is needed, it allows retrospective diagnosis using conventional blood culture bottles and enables bacteraemia to be demonstrated in 2–3 days. This technique has enabled us to demonstrate bacteraemia in 50% of cases of severe LD.

A Specimens

i Blood

At least 5 ml of blood is drawn into a sterile tube with anticoagulant (EDTA or heparin).

ii Conventional aerobic blood-culture bottles

These are often the only available samples taken before an active antibiotic treatment is initiated. We have isolated *Legionella* from aerobic blood-culture bottles which were incubated for 10 days before the attempt was made. Whether or not the composition of the broth affects the likelihood of isolating *Legionella* is not known, but it is probably not critical.

B Processing of specimens

The principle of the technique is to concentrate the legionellae which are either free or within blood monocytes, and to remove red blood cells that are inhibitory to *Legionella* (Rajagopalan & Dournon, 1986).

i Blood

The tube into which blood was taken should be centrifuged at 200*g* for 30 minutes. The plasma is then removed with a sterile pipette taking care not to remove the buffy coat that separates the plasma and red cells. The red cells and buffy coat are then transferred slowly into a 45 ml conical tube containing

35–40 ml of sterile distilled water. The mixture is vigorously vortexed, centrifuged at 200*g* for 30 minutes, and the supernate is gently removed. The pellet is then resuspended in sterile distilled water and centrifuged again.

ii Conventional blood bottles

If the blood cells have settled at the bottom of the bottles they can be aseptically removed and lysed in distilled water as described above. However, in order to recover both free and intracellular bacteria, it is better to transfer the entire broth into several conical tubes which are then centrifuged for 30 minutes at 200*g* and the resulting pellets lysed in distilled water as described above.

iii Plate inoculation

The final pellet is vortexed and 0.1 ml inoculated by streaking on to BCYE plates.

VIII Examination of culture plates

Culture plates should be examined every day in good light with the naked eye and also using a dissecting microscope.

A Interval before colonies are visible

After inoculation of BCYE medium with clinical specimens, *Legionella* will usually require at least 48 hours before growth is detectable. If specimens contain many Legionellae and are not contaminated with other bacteria, colonies can sometimes be identified after 30 hours. Growth on BCYEA is often delayed by 12–24 hours (Table 3.5).

Table 3.5 Time (in days) for colonies of *Legionella pneumophila* to become visible on BCYE and BCYEA media.

	BCYE			BCYEA		
	No.	Mean	Range	No.	Mean	Range
Endotracheal aspirates	80	3.4	1.5–9	76	4.2	2–9
TTA or BAL	24	3.3	1.5–7	20	3.9	2–6
Expectorated sputum	16	3.2	2–6	30	3.9	2–7
Total	120	3.4	1.5–9	126	4.1	2–9

Appearance of colonies may also be delayed if the patient has received appropriate antibiotics, the specimen is contaminated with other microorganisms, or a species other than *L.pneumophila* is concerned. Until recently, we routinely incubated the plates for 15 days. However, as growth has never been observed after day 9, a 10-day incubation is probably sufficient (Table 3.5).

B Appearance of colonies

i Appearance of *L.pneumophila* colonies on BCYE medium

Colonies of *L.pneumophila* have a typical 'ground-glass' texture. The younger the colonies the more obvious is this appearance. Very young colonies, seen only under the dissecting microscope, are generally green and later turn purple (Plate I.a). When older (3–5 days), they are pink–purple at the edges and whitish in their centre (Plate I.b). The ground-glass texture is 'thicker' than in very young less opaque colonies. At this stage the colonies have a well-defined edge, have a watch-glass morphology (*Legionella* colonies never have a Chinese hat morphology) and are slightly shiny (very shiny colonies are unlikely to be *Legionella*). Colonies that are more than a week old are much less characteristic, being flat and whitish. However, under the dissecting microscope a thin purple ring with discrete ground-glass appearance is generally still visible at the edge of these colonies.

ii Other *Legionella* spp.

In Europe the probability of isolating *Legionella* not belonging to the species *L.pneumophila* from clinical specimens is, as far as we know very low. Colonies of such species are less easy to recognize than *L.pneumophila*. The ground-glass appearance is less marked, they have a smooth appearance and are generally less coloured.

C Microbial colonies which may resemble those of *Legionella*

Especially on BCYEA, small colonies of *Pseudomonas* spp. and, to a lesser extent, colonies of *E.coli*, other Enterobacteriaceae and *Bacillus* spp., may very closely resemble *Legionella* colonies. In many instances the use of simple criteria can resolve this problem. A colony which morphologically resembles *Legionella* is probably not *Legionella* if:

- it appears in less than 36 hours on BCYE or in less than 48 hours on BCYEA;
- it appears adjacent to colonies of other microorganisms, especially *Candida* spp. which are often inhibitory to *Legionella*;

- on BCYEA, the colony edges are irregular (this is likely to be *Pseudomonas* spp.);
- the ground-glass pattern is concentrically arranged as on a target (the ground-glass appearance of *Legionella* colonies is irregular).

In many instances when initial differentiation is not possible re-examination of the plate a few hours later will be helpful.

IX Presumptive identification of legionellae

Colonies with the characteristic ground-glass appearance should be suspended in distilled water and kept at 4 °C (where *Legionella* survive for several months) and further identified.

A Immunofluorescence (IF)

IF is a reliable method to tentatively identify *Legionella* on an emergency basis. However, it is important to know the specificity of the reagents being used and especially if cross-reactions with *Pseudomonas* spp. occur, these can very closely mimic *Legionella* and such cross-reactions are very frequent with some commercial reagents.

In Europe, as *L.pneumophila* serogroup 1 accounts for more than 90% of culture-proven LD cases, we would recommend that reagents against this serogroup be used and any unreactive strain likely to be *Legionella* be sent to a reference laboratory.

B Growth requirements

The suspension of the putative *Legionella* is inoculated on to:

- a blood agar plate;
- a BCYE plate without L-cysteine (BCYE – Cys);
- a BCYE plate.

Bacteria that grow on any of these media other than BCYE are not *Legionella*. Care should be taken not to inoculate first the BCYE plate and then with the same loop to inoculate the other media. Transfer of even small quantities of BCYE on to the other media may allow *Legionella* to grow on these.

C Other procedures

Growth requirements and IF are sufficient for routine identification of *L.pneumophila* strains. Details of other procedures are given in Chapter 5.

Legionella isolates from clinical specimens which are not *L. pneumophila* should be sent to a reference laboratory for confirmation.

X Preparation and quality control of media

Many variants of the basic charcoal yeast-extract medium have been described. In our experience the use of buffered charcoal yeast-extract medium (BCYE) with and without antibiotics is sufficient for routine clinical work. We recommend that large amounts of the 'BCYE base' are prepared at one time as we find this is stable for several months. The various BCYE-derived media are then prepared from the base when required. The procedure used to prepare the BCYE base, common to all routine media, is indicated separately followed by the procedure to prepare the media themselves.

A Buffered charcoal yeast-extract agar (BCYE)

Base: Ingredients per litre

Yeast extract (Difco)	10.0 g
Activated charcoal (Sigma, Norit A)	2.0 g
Agar (Difco)	16.0 g
ACES buffer (Sigma)	10.0 g
1 M KOH	40.0 ml
Alpha-ketoglutarate monopotassium salt	0.25 g
Distilled water	940.0 ml

Preparation

(a) Yeast extract, charcoal and agar are mixed together.
(b) ACES buffer is added to 500 ml of distilled water warmed to 50 °C and stirred until dissolved.
(c) The KOH is added to 440 ml of distilled water.
(d) (b) and (c) are mixed together.
(e) Alpha-ketoglutarate is added to (d) and mixed well.
(f) (e) and (a) are mixed together and gently heated in a water bath (50 °C) until fully dissolved.
(g) The medium (490 ml) is poured into 500 ml bottles and autoclaved for 15 minutes at 121 °C.

The BCYE base can then be used immediately, after cooling to 50 °C, to prepare BCYE medium, BCYEA or BCYE–Cys, or stored at room temperature until needed.

BCYE medium

490 ml of BCYE base
L-Cysteine hydrochloride (Sigma)
Ferric pyrophosphate (Sigma)
Distilled water

(a) A bottle (490 ml) of BCYE base is melted and allowed to cool to 50 °C in a water bath.
(b) L-Cysteine HCl (0.4 g) is dissolved in 10 ml distilled water, filter sterilized, and 5 ml added to the BCYE base.
(c) Ferric pyrophosphate (0.25 g) is dissolved in 10 ml distilled water, filter sterilized, and 5 ml added to (b). It should be noted that the L-cysteine must be added before the ferric pyrophosphate.

The pH of the BCYE is very critical and should be 6.9 ± 0.05 at 40–45 °C. If necessary it can be corrected using KOH or HCl. One plate from each batch should be incubated at 35–37 °C for 48 hours to ensure sterility. BCYE agar plates stored at 4 °C in closed plastic bags can be kept for several weeks without obvious loss in sensitivity.

B BCYE less L-cysteine (BCYE–Cys)

This medium is identical to BCYE except that L-cysteine HCl is omitted.

C BCYE with antibiotics (BCYEA)

Ingredients per 500 ml of BCYE medium:

Polymyxin B sulphate, 40 000 IU (as base)
Vancomycin HCl, 0.250 mg
Cefamandole, 2 mg (as base)
Anisomycin, 40 mg
500 ml of BCYE medium (see above)

The antibiotics should be individually dissolved in distilled water, filter sterilized, and added to the BCYE medium. Anisomycin, which is not soluble in water at neutral pH, can be dissolved in 0.1 M HCl and then neutralized with 0.1 M NaOH. This latter anti-fungal component is very expensive and can be omitted without obvious loss of selectivity. BCYEA plates kept at 4 °C remain sensitive and selective for at least 3 weeks.

D Quality control

To ensure that each batch of medium is suitable for use one plate should be seeded with *Legionella* (about 100 cfu/0.1 ml). Colonies should be seen after 48 hours' incubation at 35–37 °C. The number of cfu should not be less than 70% of the inoculum. The test strain should not be a laboratory adapted isolate as this may grow even if the medium is not optimal. A suitable test strain can be obtained as follows:

- A guinea-pig is infected via the intraperitoneal route with 10^5 to 10^6 cfu of *L.pneumophila*.
- After 48–60 hours when the animal is obviously sick and has lost at least 10% of its body weight it is sacrificed.
- The lungs are removed aseptically (in a safety cabinet) and stored frozen in small aliquots.

E Commercially available media

We have only a limited experience of media commercially available in France. However, to date, the media produced in our laboratory have always been superior to commercial equivalents which have also shown considerable variations in quality from batch to batch. It is likely that with increasing experience the quality of commercial media will be improved.

A Laboratory Manual for *Legionella*
Edited by T. G. Harrison and A. G. Taylor

CHAPTER 4

Isolation of Legionellae from Environmental Specimens

P. J. L. Dennis

I Introduction

The first isolation of *Legionella pneumophila* from a natural habitat was from the mud of a stream (Morris *et al.*, 1979). Subsequently *L.pneumophila* and other legionellae have been isolated from rivers, streams, thermally polluted waters, natural thermal ponds, the shores of lakes (Fliermans *et al.*, 1981) and from ponds located in the blast zone of a volcano (Tison *et al.*, 1983). These studies also indicated that *L.pneumophila* is more commonly isolated from water in the temperature range 36 – 70 °C than from cooler water, with the highest isolation rates from water within the range 40 – 60 °C. Legionellae are thus aquatic bacteria which survive at temperatures that are not tolerated by most other aquatic organisms. They are not thermophiles but their tolerance of higher temperatures may give them an ecological advantage uniquely fitting them for colonizing man-made water systems which are often at higher than ambient temperatures.

Examination of water samples for the presence of legionellae has frequently been undertaken using direct immunofluorescence methods (Fliermans *et al.*, 1981). These surveys have been of limited value because of the unknown specificity of the reagents being used. The recent use of monoclonal antibodies has reduced the problems of cross-reactivity previously encountered where polyclonal antibodies were used for the examination of environmental specimens. However, bacterial isolation is still the only conclusive method of demonstrating the presence of legionellae and is required in investigations of outbreaks of LD particularly where clinical isolates may be available for comparison with environmental isolates.

L.pneumophila was first isolated by infecting guinea-pigs and fertile hens' eggs (McDade *et al.*, 1977). Subsequently it was shown that the organism could be grown on bacteriological culture media (Feeley *et al.*, 1978). Seventeen different culture media were examined and of these only Mueller – Hinton agar supplemented with 1% haemoglobin and 1% Isovitalex supported growth. On this medium colonies were only just visible after four to five days' incubation at 37 °C in 5 – 10% CO_2 in air. The medium was analysed and it was found that Isovitalex and haemoglobin could be replaced by L-cysteine hydrochloride and ferric pyrophosphate respectively. Optimal growth of *L.pneumophila* was obtained at pH 6.9 at 35 °C in air supplemented with 2.5% CO_2. An improved medium, Feeley – Gorman agar, was then developed this being rapidly superseded by charcoal yeast-extract agar (CYE) also formulated by Feeley and colleagues (1979). The CYE medium was further improved (Pasculle *et al.*, 1980) by the addition of the organic buffer ACES (*N*-2-acetamido-2-aminoethane-sulphonic acid) and α-ketoglutarate (Edelstein, 1981). Although other media have been described (Greaves, 1980; Dennis *et al.*, 1981) BCYE has become the medium of choice (Table 4.1).

Despite these improvements in culture media guinea-pig inoculation, which is both time consuming and expensive, was frequently required to isolate legionellae from many clinical and environmental samples. Such specimens are often heavily contaminated with other microflora capable of out-growing and masking the presence of legionellae on culture plates. Therefore, concurrent with the development and evolution of the base media, selective methods were sought to permit the isolation of legionellae from contaminated samples without the use of guinea-pigs.

Table 4.1 Development of culture techniques for the isolation of legionellae.

Date	Author(s)	Technique
1977	McDade, J.E., *et al.*	Guinea-pigs and eggs
1978	Feeley, J.C., *et al.*	Mueller–Hinton supplemented with 1% Hb and 1% Isovitalex
1979	Feeley, J.C., *et al.*	Charcoal yeast-extract agar (CYE)
1980	Pasculle, A.W., *et al.*	CYE and ACES buffer
1980	Greaves, P.W.	Blood agar with ferric pyrophosphate and cysteine
1981	Edelstein, P.H.	CYE and α-ketoglutarate (BCYE)
1981	Dennis, P.J., *et al.*	*Legionella* blood agar

Various combinations of antimicrobial agents have been used to make the basal medium selective (Table 4.2). For example, Edelstein and Finegold (1979) used polymyxin B and vancomycin, Greaves (1980) a combination of vancomycin, trimethoprim, colistin and amphotericin B, and Bopp and colleagues (1981) cephalothin, colistin, vancomycin and cycloheximide. Two such selective media proved to be particularly useful: BMPAα (Edelstein, 1981) where cefamandole, polymyxin B and anisomycin are added to BCYE, and GVP (Wadowsky & Yee, 1981) in which a mixture of glycine, vancomycin and polymyxin B is used. Subsequently GVP medium was modified by the addition of ACES buffer, α-ketoglutarate, anisomycin and by reducing the concentration of vancomycin to 1 μg/ml (Edelstein, 1982). This modified Wadowsky and Yee medium (MWY) is commercially available and has been widely used. More recently GVP was modified by the addition of cycloheximide, α-ketoglutarate and the reduction of the vancomycin to 1 μg/ml. This improved medium (GVPG) is used in the UK (Dennis *et al.*, 1984a) and described in detail below.

Attempts have also been made to enhance the coloration of legionella colonies on BCYE agar by the addition of bromocresol-purple and bromothymol-blue, to permit differentiation between species (Vickers *et al.*, 1981). The value of adding these dyes to selective media is, however, somewhat doubtful.

The observation, initially made by Wang and colleagues (1979), that legionellae could survive exposure to acid conditions (pH 2.0) for short periods led to the development of several acid pretreatment methods (Greaves, 1980; Bopp *et al.*, 1981). In addition it was also shown that, relative to many other aquatic bacteria, legionellae are tolerant to heat (Muller, 1981; Hernandez *et al.*, 1983) withstanding temperatures of 50 °C and above for a considerable time (> 30 min) (Dennis *et al.*, 1984b). Consequently simple acid and heat pretreatment methods have been developed to reduce the numbers of non-legionellae in samples, thus increasing the probability of isolating the legionellae. When compared with the isolation rate obtained using guinea-pigs, the use of either BMPAα or GVPC together with heat and acid pretreatments results in a significant improvement in sensitivity (Dennis *et al.*, 1984a).

Table 4.2 Development of selective culture media for the isolation of *Legionella* species.

Date	Author(s)	Medium
1979	Edelstein, P.H., and Finegold, S.M.	CYE plus polymyxin B and vancomycin
1980	Greaves, P.W.	Blood agar (supplemented) vancomycin, trimethoprim, colistin and amphotericin B
1981	Edelstein, P.H.	BCYE plus cefamandole, polymyxin B and anisomycin (BMPAα)
1981	Bopp, C.A., *et al.*	CYE plus cephalothin, colistin, vancomycin and cycloheximide
1981	Wadowsky, R.M., and Yee, R.B.	CYE plus glycine, vancomycin and polymyxin B (GVP)
1982	Edelstein, P.H.	BCYE plus glycine, vancomycin (1 μg/ml), polymyxin B and anisomycin (MWY)
1984a	Dennis, P.J., *et al.*	BCYE plus glycine, vancomycin (1 μg/ml), polymyxin and cycloheximide (GVPC)

At present the most efficient *in vitro* method used in this laboratory for the isolation of legionellae is the combination of GVPC media together with acid or heat pretreatments (Dennis *et al.*, 1984a). The choice of media and pretreatment method is not dependent on the site from which a sample is collected, e.g. cooling tower or hot domestic water system, although different microbial populations may occur in these samples. As illustrated by the data in Figure 4.1 the methods outlined above are equally effective for all types of sample. This combination of pretreatment and selective media formed part of the isolation protocol used in a Public Health Laboratory Service (PHLS) survey of hospital and hotel water systems for legionellae (Bartlett *et al.*, 1983) and will be described in detail. However, before isolation can be attempted a suitable water sample likely to contain the organism must be collected.

Samples of water are taken and examined for the presence of legionellae for many different reasons. The most obvious reason is to assist the epidemiological studies during the investigation of a building epidemiologically implicated in an outbreak of LD. In buildings not associated with disease other factors influence the decision as to whether the water should be tested. For example it may be thought appropriate to test for *Legionella* (and other opportunistic pathogens, e.g.

pseudomonads) in buildings which house patients undergoing immunosuppressive or steroid therapy who are therefore particularly susceptible to infection. However, the need to examine water systems in buildings housing healthy individuals, e.g. in a maternity unit, is questionable.

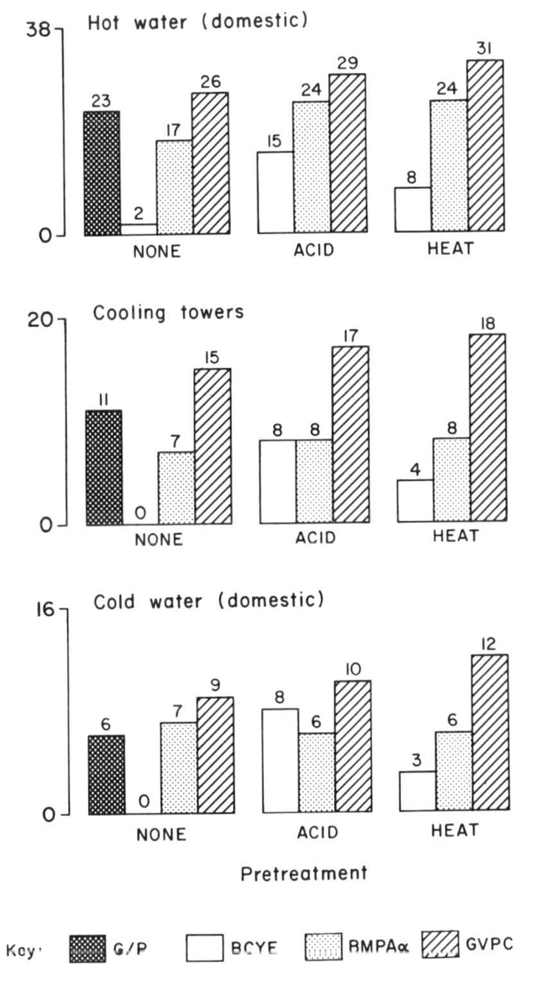

Figure 4.1 Comparison of legionellae isolation from hot, cold and cooling-tower water samples, using either guinea-pigs or selective media with or without pretreatment. G/P = guinea-pigs: BCYE, BMPAα, and GVPC are described in the text.

The publicity given to cases of LD has led to attempts to improve the quality of water in buildings and to improve the microbiological standards of cooling towers. The responsibility for such improvements is that of plant engineers or maintenance personnel, who may request culture of specimens to show that the measures undertaken to eradicate or control the organism were effective.

Laboratory reports on the culture of environmental specimens are only useful if sampling is undertaken in a methodical and logical manner. In this way sampling will not only provide evidence that a water system is colonized with legionellae but should also indicate where within the system the organism is present in greatest numbers. Such information is important regardless of whether the investigation is undertaken as a result of an outbreak or as a routine preventative examination. Knowledge of where within a water system the bacteria are located is essential if measures undertaken to eradicate or control them are to be effective.

The examination of a water system on a single occasion provides little useful information. Fluctuations in temperature, changes in water chemistry or water use will influence the microbial populations within the system which therefore must be examined at regular intervals to take account of such fluctuations. Sampling frequency will be influenced by the size and complexity of the water system, its use, the degree of colonization and the type of control procedure used. In the hospital environment the susceptibility of patients to infection with opportunistic pathogens may also dictate the sampling frequency. The influence of these variables may be determined by preparation of microbiological standards (baseline data) for each water system. Significant deviations from these standards would indicate defective equipment or control measures. Establishment of this baseline may take several years. However, once the baseline is established the frequency of sampling required to monitor the system can be reduced.

II Sample collection

A Water services within buildings

Detailed discussion of the protocols used for sample collection is outside the scope of this contribution therefore only a brief guide is given here.

When investigating the water services within a building, to determine whether they are colonized with legionellae, it is essential to prepare or obtain a simple schematic diagram of these services. The following features, if present, should be noted:

1. The location of the incoming mains supply and/or private source.
2. The location of cisterns, booster vessels and pumps.
3. The location of calorifiers, water heaters.

4. The type of fittings used in the system, e.g. taps, showers, valves, and the material from which the pipework is constructed.
5. Whether or not cooling towers or heating circuits are present.
6. Whether air conditioning systems and humidifiers are present within the building.

The route of the services should be traced from the point of entry of the water supply. The condition of pipes, the jointing methods used, the presence of lagging, sources of heat, and the standard of protection afforded cisterns should be noted on the diagram. A careful note should also be made of disconnected fittings, 'dead-ends', and cross-connections with other services.

Having identified these sites, water samples should be taken from:

- the incoming supply;
- cisterns and calorifiers;
- an outlet close to, but downstream of, each cistern and calorifier;
- the distal point of each service;
- the water entering and leaving any fitting under particular suspicion.

Samples should be collected in suitable sterile containers and measurements made in the order listed below. It is important that the outlet is not flushed before samples are collected. The external surface and rim of the outlet being sampled should be clean and free from deposits.

1. A one-litre 'preflush' sample of water is collected.
2. The residual chlorine levels are measured (if required).
3. A thermometer is placed in the water flow and the outlet is allowed to discharge until the temperature stabilizes. The initial and final temperatures, and the time taken to reach the latter, are recorded.
4. A five-litre 'after-flush' sample of water is collected.

The initial one-litre sample is intended to indicate the level of contamination at the sample point/fitting and the final five-litre sample should reveal the quality of the water being supplied to the fitting.

B Cooling towers

Cooling towers, using water as the coolant, vary markedly in their size and design. In some, air is drawn through the tower by fans mounted at the top (induced draught), while in others fans mounted at the bottom direct air up through the tower (forced draught). In both of these designs water flows down the tower through the pack or fill with the air moving up the tower against the water flow to effect cooling. These are generally termed induced or forced draught counterflow towers. Other towers may induce or force air at right angles to the water flow and are termed induced or forced draught crossflow towers. The size of the system and the volume of water within it can vary from less than a tonne to tens of tonnes of water coolant, according to the heat load they are designed to dissipate. Irrespective of the design all will have a pond, sparge pipes and condenser units from which samples may be taken (Figure 4.2).

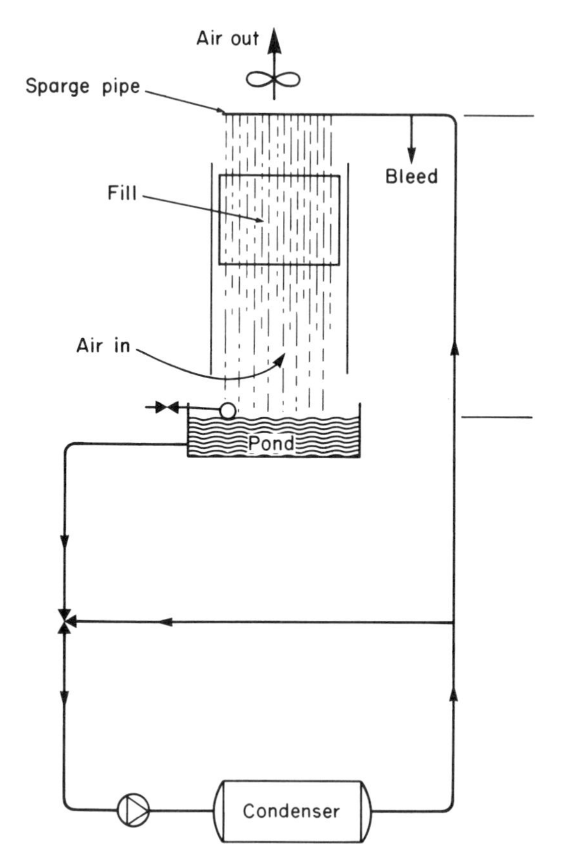

Figure 4.2 Schematic diagram of an induced-draught cooling tower.

A simple diagram showing the cooling tower and the services to which it is connected should be prepared or obtained. This should show cross-connections between towers, the extent of the pipework, and the working volume of the system. A note should be made of the condition of the tower and the presence and extent of biofouling. The following information should be sought; whether or not descalents, biodispersants and biocides are used, their proprietary names, the concentrations at which they are used, the site at which they are added to the system, and the frequency of their addition. A knowledge of the volume of the water system and the quantity of biocide added to the tower will allow a check to be made as to whether the chemicals were being used according to the manufacturers' recommendations.

Water samples should be taken from the incoming supply to the tower, either from the header tank or the ball valve located in the tower pond. Samples should also be taken of the pond water furthest from the water make-up, and of the water returning from the circulation system at the point of entry to the tower. For both building services and cooling towers, sludge, swabs of shower-heads, pipes and taps or other material should be collected and hydrated either with water taken from the sampling site or with sterile tap water.

In addition to the general sampling points any fitting thought, from epidemiological considerations, to be a source of infection should be located and water entering and leaving it sampled. Legionellae have been isolated from jet nebulizers and portable room humidifiers (Arnow *et al.*, 1982) and such equipment has been implicated as the source for a number of infections. Therefore, any equipment which is likely to contain water or produce aerosols, and which may come into contact with susceptible individuals, should also be sampled. The protocols for sampling such equipment have been detailed by Ashworth and Colbourne (1987).

III Concentration of legionellae from water samples

Before the advent of selective media, primary isolation of legionellae depended upon inoculation of guinea-pigs (McDade *et al.*, 1977), and at least 10^6 cells were required to reliably achieve isolation (Morris *et al.*, 1979). In consequence concentration was required in the case of water samples containing low numbers of bacteria. To determine the volume of water that had to be concentrated, samples were graded according to the number of cells seen by immunofluorescence using antibody prepared against legionellae. The fewer bacteria seen the greater volume of water required. Although not essential using *in vitro* culture methods, it is still desirable to concentrate the bacteria in water samples before attempting to isolate them. Two methods of concentration can be used, either centrifugation or filtration.

Samples can be centrifuged using conventional rotors but sample volume is then restricted by the size of the centrifuge buckets. Alternatively continuous centrifugation may be used but few laboratories have access to the appropriate equipment and only one or two samples can be processed in a day. The second method of concentration is by filtration using membranes of small pore size to retain the bacteria required for culture. Filter membranes of 0.22 μm pore size are the most suitable. Ease of use and the possibility of processing large volumes of water rapidly, make filtration the method of choice and it is described in detail here.

The water samples can be filtered through membranes of 47 mm, 90 mm or 142 mm diameter depending upon the clarity and volume of the water being examined. Nylon membranes of 142 mm diameter and 0.22 μm pore size are the most suitable for filtering up to five litres of water. Nylon membranes are stronger than cellulose – ester or polycarbonate membranes and able to withstand disinfection with recirculated boiling water, are easily handled, may be used at high flow rates and bacteria are easily eluted from them. The use of a 0.22 μm pore size allows more legionellae to be retained than does a 0.45 μm pore size membrane. However, samples which are cloudy or contain particulate matter should be first filtered through a coarse grade filter. After filtration the membranes are placed in a measured volume (50 ml) of filtrate in a suitable container and shaken vigorously until the deposit on the membrane is resuspended. If the original sample is thought to contain bacterial inhibitors (e.g. biocides) material on the membrane should be resuspended in sterile chlorine-free tap water, in place of the filtrate. The concentrated sample is then used as the inoculum.

When using large 142 mm diameter filter assemblies to filter sample volumes of five or more litres, it is impractical either to pressurize the sample vessel, or to draw water through the membrane by vacuum. A simple, effective and safe alternative is to draw the water from the sample vessel and then to pump the water through the filter assembly using a peristaltic pump. This method can also be adapted to 'disinfect' the assembled apparatus by using the pump to pass boiling water through it, the hot filtrate being returned to the boiling vessel (Dennis *et al.*, 1981). Water at 100 °C pumped through the assembly in this manner for approximately five minutes will eradicate legionellae and most other vegetative organisms. It is essential that the water remains near boiling throughout the process. Any additional ports, tubing or pressure relief valves must also be flushed with boiling water to prevent cross-contamination when filtering subsequent water samples. The assembly can then be left to cool, or cold sterile distilled water can be pumped through to waste before the next sample is processed.

IV Culture

A Selective media

Selective media agar plates (either GVPC or MWY) are inoculated with 0.1 ml volumes of the concentrated sample which are spread over the entire surface of the plates using sterile bent-glass rods or 'spreaders'. For each sample a further two sets of selective plates are inoculated; one after pretreating a portion of the concentrated sample with heat, and the other set inoculated with 0.1 ml of the sample after acid pretreatment (Bopp *et al.*, 1981).

B Heat pretreatment

1. 10 ml of the sample concentrate is taken and placed in a 50 °C water bath for 30 minutes.
2. The heated sample is mixed well and 0.1 ml spread on to a selective media plate using a sterile bent-glass rod or 'spreader'.
3. The culture plate is incubated after allowing the inoculum to absorb.

C Acid pretreatment

Buffer

3.9 ml of 1.2 M HCl
25 ml of 0.2 M KCl
Adjust to pH 2.2 using 1 M KOH

Procedure

1. Concentrated sample (10 ml) is taken and centrifuged in sealed buckets at approximately 1000*g* for 20 minutes.
2. The supernatant is removed leaving approximately 1 ml of fluid.
3. HCl–KCl buffer (9 ml) is added and the mixture gently shaken for five minutes.
4. The treated sample (0.1 ml) is spread on to BCYE and selective media using a sterile bent-glass rod or 'spreader'.
5. The culture plates are incubated after allowing the inoculum to absorb.

A combination of the selective media and pretreatment techniques described above provides an efficient and effective means of isolating legionellae from water, which can be used on samples gathered from widely differing sources (Dennis *et al.*, 1984a).

D Incubation and examination of culture plates

Plates can be incubated in air, or preferably in air with 2.5% CO_2, at 35–37 °C for up to 10 days. Due to the extended incubation period, humidification is necessary to prevent desiccation of the culture plates. In incubators not equipped with humidifiers a tray of water should be placed at the base of the incubator. If this is impractical, or undesirable, plates can be incubated in anaerobic jars with the valves removed and a water-soaked absorbent pad placed in the bottom.

It is unnecessary to examine plates before the third day post-inoculation. Colonies of legionellae can take between 48 and 72 hours to appear and some may not be visible for five or more days. It is therefore advisable to incubate plates for up to 10 days. As other bacteria commonly grow even on the selective media it is important to be able to recognize *Legionella* colonies. These have a grey–blue–purple or sometimes lime-green coloration on GVPC or BCYE. The colonies are smooth with an entire edge. When viewed under a plate microscope, colonies have a 'ground-glass' appearance (Plate I.a,c). Plates should also be examined under ultraviolet light (366 nm). Colonies of *L.bozemanii*, *L.gormanii*, *L.dumoffii*, *L.anisa*, *L.cherrii* and *L.parisiensis* fluoresce a brilliant white, while *L.rubrilucens* and *L.erythra* appear red (Plate II.h). The colonies of *L.pneumophila* appear dull green or yellow. This autofluorescence can be an aid to identification when a sample contains strains of several species of *Legionella*. Putative *Legionella* colonies are subcultured on to BCYE and on to BCYE from which the L-cysteine HCl has been omitted (BCYE–Cys) or blood agar.

V Identification

Isolates that fail to grow on BCYE–Cys or blood agar and are small, poorly staining, Gram-negative rods, are presumptively identified as legionellae. A limited number of biochemical tests are available and can be used to assist identification, but they cannot be generally relied upon for the differentiation of species. Specific identification can be achieved by immunofluorescence or slide agglutination, using specific antisera (Thacker *et al.*, 1985a) together with the determination of cellular fatty acid composition and isoprenoid quinone profiles (Moss *et al.*, 1977; Collins & Gilbart, 1983). However, occasionally identification can only be achieved by the determination of nucleic acid homology. Further details with regard to identification are described in Chapters 5–7.

VI Preparation of media

A Buffered charcoal yeast-extract medium (BCYE)

Ingredients (per litre)

Yeast extract (Oxoid)	10.0 g
Agar (Oxoid)	12.0 g
Activated charcoal (Norit SG)	1.5 g
α-Ketoglutarate, monopotassium salt	1.0 g
ACES* buffer (Sigma)	10.0 g
KOH pellets	2.8 g
L-Cysteine	0.40 g
Ferric pyrophosphate	0.25 g
Distilled water to 1 litre	

* ACES buffer is *N*-2-acetamido-2-aminoethane-sulphonic acid. Its pK_a is 6.9 at 20 °C.

Preparation

It is essential that the ACES buffer be hydrated in the following manner to avoid possible denaturation of the yeast extract. ACES buffer is added to 500 ml of distilled water and dissolved by placing in a water bath (45 – 50 °C). This solution is then mixed with 480 ml sterile distilled water in which the KOH pellets have been dissolved. This mixture is used to hydrate the activated charcoal, yeast extract, monopotassium α-ketoglutarate and agar. The medium is then autoclaved at 121 °C for 15 minutes, cooled to 50 °C and membrane-filter sterilized solutions of L-cysteine HCl (0.40 g in 10 ml distilled water) and soluble ferric pyrophosphate (0.25 g in 10 ml distilled water) are added. The pH of the final medium should be 6.9 and can be adjusted with sterile 0.1 M KOH or 0.1 M H_2SO_4.

B Buffered charcoal yeast-extract less L-cysteine (BCYE – Cys)

This medium is identical to BCYE medium except the L-cysteine is omitted.

C Antibiotic supplemented selective media

GVPC (Glycine, Vancomycin, Polymyxin, Cycloheximide) medium is a modification of that described by Wadowsky and Yee (1981) and is prepared as follows:

- *Glycine* Ammonium free glycine (3 g/l) is added to BCYE base before autoclaving and after the rehydration of the ACES and the addition of α-ketoglutarate to the base medium.
- *Polymyxin B* 200 mg of polymyxin B sulphate is added to 100 ml distilled water, filter sterilized, dispensed in 5.5 ml quantitics, and stored at –20 °C. One volume (5.5 ml) is added to a litre of BCYE base to give a final concentration of 79 200 IU/l.
- *Vancomycin* 100 mg of vancomycin hydrochloride is added to 20 ml distilled water, filter sterilized, dispensed in 1 ml quantities, and stored at –20 °C. One volume (1 ml) is added to a litre of BCYE base ot give a final concentration of 5 mg/l.
- *Cycloheximide* 2 g of cycloheximide is dissolved in 100 ml distilled water, sterilized as above, dispensed in 4 ml volumes, and stored at –20 °C. One volume is added to a litre of BCYE to give a final concentration of 80 mg/l. Cycloheximide is extremely toxic and so the recommended safety precautions should be observed when handling this chemical.

VII Quality control of media

It should be remembered that the BCYE medium and its additives are heat sensitive. Prolonged heating during normal sterilization procedures, or heating at too high a temperature can severely affect the nutritional qualities of the medium. Batch-to-batch variation of the ingredients of the medium (e.g. α-ketoglutarate) can also severely affect its performance. It is therefore essential that the quality of each newly prepared batch of medium should be assessed for its ability to promote the growth of *L.pneumophila* within three days.

It is usual in general bacteriology to assess media using strains of bacteria previously isolated in the laboratory and stored either on slopes or frozen in broth. In the case of *Legionella* isolation the use of such strains can lead to false assumptions concerning the quality of the media under test. Legionellae become easily adapted to growth under laboratory conditions and can then be cultured on media that will not support the primary isolation of 'wild' strains. Ideally BCYE media should be tested using *L.pneumophila* strains that previously have not been grown *in vitro*. The following protocol can be used to assess the quality of BCYE agar and selective media.

BCYE media should be inoculated with a measured volume of macerated spleen taken from a guinea-pig previously infected with *L.pneumophila*. Several different dilutions of the spleen tissue can be used and the numbers of colonies obtained compared with those recorded for previous batches together with the time taken for colonies to become visible and their morphology discernible. Simple statistical tests can also be applied to determine if there is any significant difference in the colony counts between batches.

In the case of selective media water samples known to contain legionellae are very useful for quality control purposes. Plates are inoculated with measured volumes of the sample. The bacterial growth can be assessed as above, and the effectiveness of the selective supplements also ascertained.

It is advisable to maintain records of the quality control tests and to note when new batches of media (commercial or otherwise) or their ingredients are used so that any inadequacies detected may be rectified.

It is not possible or prudent to recommend any particular manufacturer of BCYE media or the materials used to make it, as batch-to-batch variation can occur. It is recommended that a number of products from different suppliers are tested so that suitable batches of media or the ingredients can be identified and purchased.

A Laboratory Manual for *Legionella*
Edited by T. G. Harrison and A. G. Taylor

CHAPTER 5

Phenotypic Characteristics of Legionellae

T. G. Harrison and A. G. Taylor

I Introduction

The generally accepted classification of the family Legionellaceae is based on the work of Brenner and colleagues (see Brenner *et al.*, 1984). These workers take the definition of a 'genetic' species as a group of strains whose DNAs are 70% or more related at optimal reassociation conditions; 55% or more related at stringent conditions and have 6% or less divergence in their related sequences. Further they argue that a genus is a man-made division and should consist of phenotypically similar species. For this reason they oppose the creation of additional genera within the family Legionellaceae (Brenner *et al.*, 1985). This viewpoint is not shared by all workers and it has been proposed that the family should contain three genera; *Legionella*, *Tatlockia* and *Fluoribacter* (Garrity *et al.*, 1980). Although not widely used these alternative names are sometimes used in the literature. Throughout this chapter the classification as proposed by Brenner and colleagues is adhered to and therefore at the time of writing there are 51 serogroups comprising 34 species within a single genus *Legionella* (Brenner, 1986; Appendix 1). Some species can only be identified by DNA relatedness studies and cannot be separated by phenotypic or serological tests.

Legionella are Gram-negative rods 0.3 – 0.9 μm in width and 2 – 20 μm or more in length. They are usually poorly motile with one or two polar flagella although non-motile strains do occur. Branched chain fatty acids predominate in the cell wall which also contains major amounts of ubiquinones with more than ten isoprene units in the side chain. L-Cysteine and iron salts are required for growth *in vitro*, carbohydrates are not fermented or oxidized and nitrate is not reduced to nitrite. *Legionella* are urease negative, catalase positive and give variable results in the oxidase test.

Multiple strains of *Legionella* are sometimes isolated on the same culture plate from a single environmental or clinical specimen. These may be strains of different species, serogroups or subgroups. Before characterizing any *Legionella* isolate it is essential to ensure that the culture is pure.

The phenotypic characteristics and tests described below are intended to enable the identity of putative legionellae to be established at least to the genus level and in some cases to species level. Used in association with serological methods most species can be correctly identified.

II Identification of legionellae

A Nutritional requirements

Legionellae are most readily distinguished from similar organisms by their nutritional requirement for L-cysteine and, to a lesser extent, iron salts. Thus a catalase-positive Gram-negative organism, which grows on BCYE but fails to grow on either BCYE without L-cysteine (BCYE – Cys) or blood agar (BA), can tentatively be identified as a legionella (Plate II.f).

Although all legionellae require L-cysteine for primary isolation, *L.oakridgensis* is reported to grow in its absence thereafter (Orrison *et al.*, 1983b). The type strain (NCTC 11531) grows well on BCYE – Cys in our laboratory as will the type strain of *L.spiritensis* (NCTC 11990). Similarly some highly adapted laboratory strains of other species will show slight growth in the absence of L-cysteine. For the purposes of identification an isolate is considered not to grow on BCYE – Cys if no colonies are visible after 72 hours' incubation at 37 °C. After prolonged incubation, many isolates give scanty growth at the point where the initial 'streak' was made. This is possibly due to the carry-over of small quantities of L-cysteine from the BCYE plate from which the inoculum was taken.

It should be noted that *Francisella tularensis* also requires L-cysteine for growth and so will satisfy the above criteria. Although this organism is not likely to be encountered in the UK, caution should be exercised in areas where tularaemia is endemic as this highly infectious pathogen has been confused with *Legionella* (Macleod *et al.*, 1984). Thacker and colleagues (1981) reported a second source of possible confusion. During the course of routine culturing of human lung tissue, these workers isolated six Gram-negative organisms which resembled legionellae by both their growth requirements and their cell-wall fatty-acid profiles. However, these bacilli were distinguishable from legionellae as they were thermophilic and sporeforming. The authors concluded that in all probability these organisms were contaminants derived from the batch of culture media being used.

B Morphology of colonies

Colonies of putative *Legionella* spp. should be examined using a dissecting microscope. *Legionella* colonies are convex, have an entire edge, glisten and have a characteristic granular or 'ground-glass' appearance which is most pronounced in young colonies (Plate I.c). Coloration varies from a blue/green when the colonies are first visible (Plate I.a) becoming pink/purple as they grow larger. As the colonies age they become less characteristic being white/grey and smoother, however the pink/purple coloration can still be seen at their edges (Plate I.b). As the name suggests the colonies of *L.erythra* have a slight red coloration (Plate II.g).

C Gram's stain

Legionellae can be satisfactorily stained by Gram's method providing a suitable counterstain is used. Either 0.5% (w/v) safranin for 5 – 10 minutes or 0.1% (w/v) basic fuchsin for 30 seconds is appropriate. Legionellae typically appear as slender Gram-negative rods. Depending on the age of the culture examined, the bacteria vary morphologically from coccal rods 2.0 μm in length to highly filamentous forms in excess of 20 μm, the latter predominating in older cultures (Plate I.d).

D Catalase reaction

All *Legionella* give a positive reaction in the catalase test. The procedure should be performed in a test tube in compliance with the Howie Code of Practice (1978). Using a wooden stick, a colony is taken from a 48–72 hour BCYE plate and emulsified in 3% H_2O_2. Bubbles of oxygen will rapidly form at the tip of the stick in a positive test. With some strains (particularly *L.pneumophila*) the reaction is rather weak taking 5–10 seconds for bubbles to form. In the case of *L.pneumophila* this is probably because they do not possess a catalase but rather a peroxidase which nevertheless can catalyse the breakdown of H_2O_2 thus giving a positive result (Pine *et al.*, 1984).

E Nitrate reduction

Legionellae fail to reduce nitrate to nitrite and this feature can be helpful in differentiating them from members of the Enterobacteriaceae. This can easily be tested by heavily inoculating a BCYE slope containing 0.2% KNO_3. After 48 hours' incubation at 37 °C 0.5 ml of solution A is added, followed by 0.5 ml of solution B. The test is positive if a red colour develops. If colour does not form zinc powder should be added, residual nitrate will then turn the solution red confirming the nitrate test to be negative.

Solution A 0.8% (w/v) sulphanilic acid in 5 N acetic acid (dissolve by gently heating).

Solution B 0.5% (w/v) *N,N*-dimethyl-1-naphthylamine in 5 N acetic acid (dissolve by gently heating).

Store in the dark at 4 °C.

A positive control strain (e.g. *Escherichia coli*) and an uninoculated agar slope should be included with each test. All *Legionella* species give a negative result in this test.

F Flagellar staining

With the exception of *L.oakridgensis* all species of *Legionella* are motile, but only weakly so, having one or two polar flagella. The flagella from all species share common antigens and can be visualized using a Legionellaceae flagellar antiserum (see page 66). Alternatively the flagella can be demonstrated by one of the many described 'silver-plating stains', based on Fontana's stain for spirochaetes. The method detailed below is simple, rapid and has provided satisfactory results in this laboratory. Dependent on the 'passage' history of the isolate being examined and the growth conditions used flagella may not be present. In our experience this is rarely the case with recently isolated stains.

The method given here is a modification of Rhodes 'silver-plating stain' as described by West and colleagues (1977).

Reagent A (mordant)

25 ml of saturated aluminum potassium sulphate
50 ml of fresh 10% (w/v) tannic acid
5 ml of 5% (w/v) ferric chloride
Add together in order, a black mixture will result

Reagent B (silver stain)

90 ml of 5% (w/v) silver nitrate
Concentrated ammonium hydroxide is slowly added until a brown precipitate is formed. Addition of ammonium hydroxide is continued dropwise until the precipitate just redissolves. The resulting solution should be slightly cloudy and remain so when agitated.

Both reagents are stable for over six months if stored in the dark at 4 °C.

Procedure

A small amount of growth is taken from a 48–72 hour culture and gently emulsified in distilled water to give a faintly cloudy suspension. A drop of the suspension is applied to the top of a clean microscope slide and allowed to run to the bottom; the smear is then left to air-dry. Each smear is covered with Reagent A for 4 minutes before being gently but thoroughly rinsed in distilled water. Reagent B is then added and heated until steam forms. The slide is left for 4 minutes before being rinsed with distilled water and dried for viewing. The typical appearance of legionellae revealed by this stain is shown in Plate II.e. It should be noted that excessive background silver deposit will result if dirty slides are used.

III Phenotypic characteristics helpful in the speciation of legionellae

There are a number of simply demonstrated phenotypic characteristics which allow the Legionellaceae to be subdivided into groups of species, or in some cases into individual species. Such results are summarized in Table 5.2. It should be remembered that these characteristics have been determined using small numbers of strains of each species, and in some cases only a single isolate has been examined. The reliability of these phenotypic markers is therefore uncertain and identification of an isolate should always be confirmed by cell-wall chemistry or serological studies.

A Hippurate hydrolysis

This test is used to determine whether an organism can hydrolyse sodium hippurate to benzoic acid and glycine, a characteristic common to many bacterial species. The procedure used for legionellae was first described by Hébert (1981) and requires overnight incubation of the strain in aqueous sodium hippurate. Glycine is then detected using a ninhydrin reagent. The test is used to differentiate *L.pneumophila* from all other *Legionella* spp.

Reagents

Sodium hippurate
A 1% (w/v) solution of sodium hippurate is prepared in distilled water. The solution is sterilized by membrane filtration (0.2 μm), dispensed in 0.4 ml volumes into screw-cap bottles (bijoux) and stored frozen at −20 °C until required.

Ninhydrin solution
A 3.5% (w/v) solution of ninhydrin is prepared in a 1:1 (v/v) mixture of acetone and n-butanol. The solution is stored in the dark at room temperature and will keep indefinitely.

Procedure

A bottle of 1% sodium hippurate is thawed and brought to 37 °C. A loopful of growth from a 48−72 hour incubated BCYE agar plate is emulsified in the solution to give a dense suspension and placed at 37 °C for 18−20 hours. After incubation, 0.2 ml of ninhydrin solution is added, thoroughly mixed, and the bottle placed at 37 °C for 10 minutes. The bottles are removed and observed for colour development within a further 20 minutes at room temperature. All shades of purple to blue are considered positive (a very pale purple being considered weakly positive) and shades of grey and pale yellow are negative.

Controls strains
A known positive strain (e.g. *L.pneumophila* Knoxville-1, NCTC 11286) and a known negative strain (e.g. *L.pneumophila* Los Angeles-1, NCTC 11233), together with an uninoculated bottle of 1% sodium hippurate should be processed each time the test is performed.

Interpretation

L.pneumophila strains will, with a few rare exceptions, hydrolyse hippurate. All other species, except for perhaps *L.feeleii* about which there is some confusion, are negative in this test. The two noted *L.pneumophila* strains which give negative results are Los Angeles-1 (NCTC 11233), the type strain for serogroup 4, and San Francisco-6, also a serogroup 4 strain (Hébert, 1981).

In the original description of the species *L.feeleii*, the type strain WO-44 was reported to hydrolyse hippurate (Herwaldt *et al.*, 1984). However, this was subsequently disputed by Brenner and colleagues (1985) who found this strain to be negative. The type strains of *L.feeleii* serogroup 1 (NCTC 12022) and serogroup 2 (NCTC 11978) are both negative in our laboratory. The single isolate of *L.spiritensis* (NCTC 11990) gives a very weakly-positive reaction.

Thus hippurate hydrolysis is a simple method of differentiating *L.pneumophila* from the other *Legionella* species.

B Autofluorescence

The exhibition of autofluorescence by some species of legionellae when exposed to long wavelength ultraviolet light is a rapid and useful technique for the differentiation of *Legionella* species into three groups (Table 5.1). Some species which are difficult to separate serologically can be readily distinguished in this way (e.g. *L.bozemanii* serogroup 2 from *L.longbeachae* serogroup 2).

Autofluorescence is seen when the organism is grown on BCYE or a clear media such as TBYE (see below). As soon as colonies are visible the plates are examined

Table 5.1 Autofluorescence of *Legionella* species examined under long wavelength UV.

No fluorescence	Blue/white fluorescence	Red fluorescence
L.pneumophila	*L.anisa* *	*L.erythra*
L.feeleii	*L.bozemanii*	*L.rubrilucens*
'*L.geestiae*'	*L.cherrii*	
L.hackeliae	*L.dumoffii*	
L.israelensis	*L.gormanii*	
L.jamestowniensis	*L.parisiensis*	
L.jordanis	*L.steigerwaltii*	
'*L.londoniensis*'		
L.longbeachae		
L.maceachernii		
L.micdadei		
'*L.nautarum*'		
L.oakridgensis		
'*L.quateriensis*'		
L.sainthelensi		
L.santicrucis		
L.spiritensis		
L.wadsworthii		
'*L.worsliensis*'		

* In the original description of *L.anisa* one of the five isolates failed to autofluoresce (Gorman *et al.*, 1985).

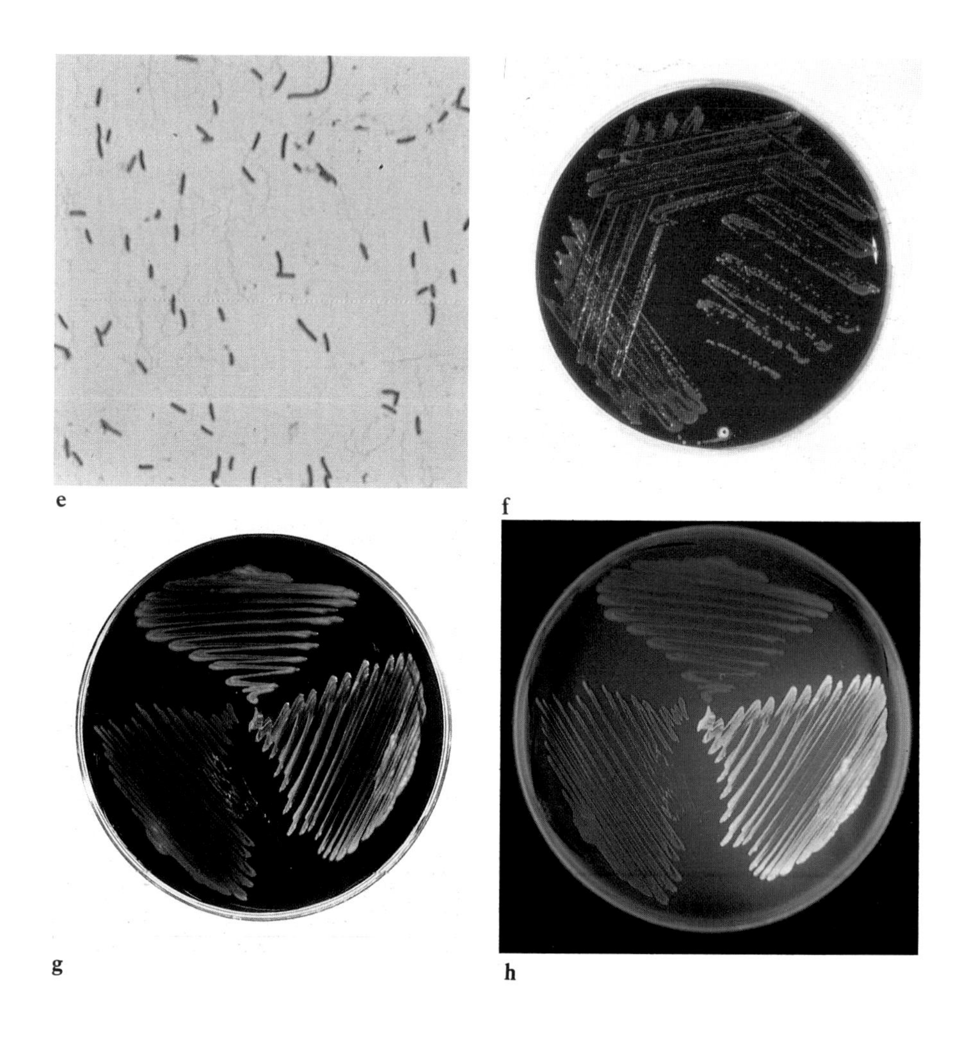

Plate II

e. Silver-stained *L.pneumophila*, taken from BCYE agar after 48 hours' incubation, showing the presence of polar flagella (original magnification: ×1000).

f. A pure culture of *L.pneumophila* on BCYE agar after 72 hours' incubation. Note the organism was recovered from a glass bead (also shown) which had been stored in liquid nitrogen as described in Appendix 2.

g, h. Demonstration of autofluorescence. Appearance of *L.bozemanii*, *L.erythra*, and *L.pneumophila* on BCYE when examined: (g) under natural light, and (h) under long wavelength UV light. Note the slight reddish pigmentation of *L.erythra* growth.

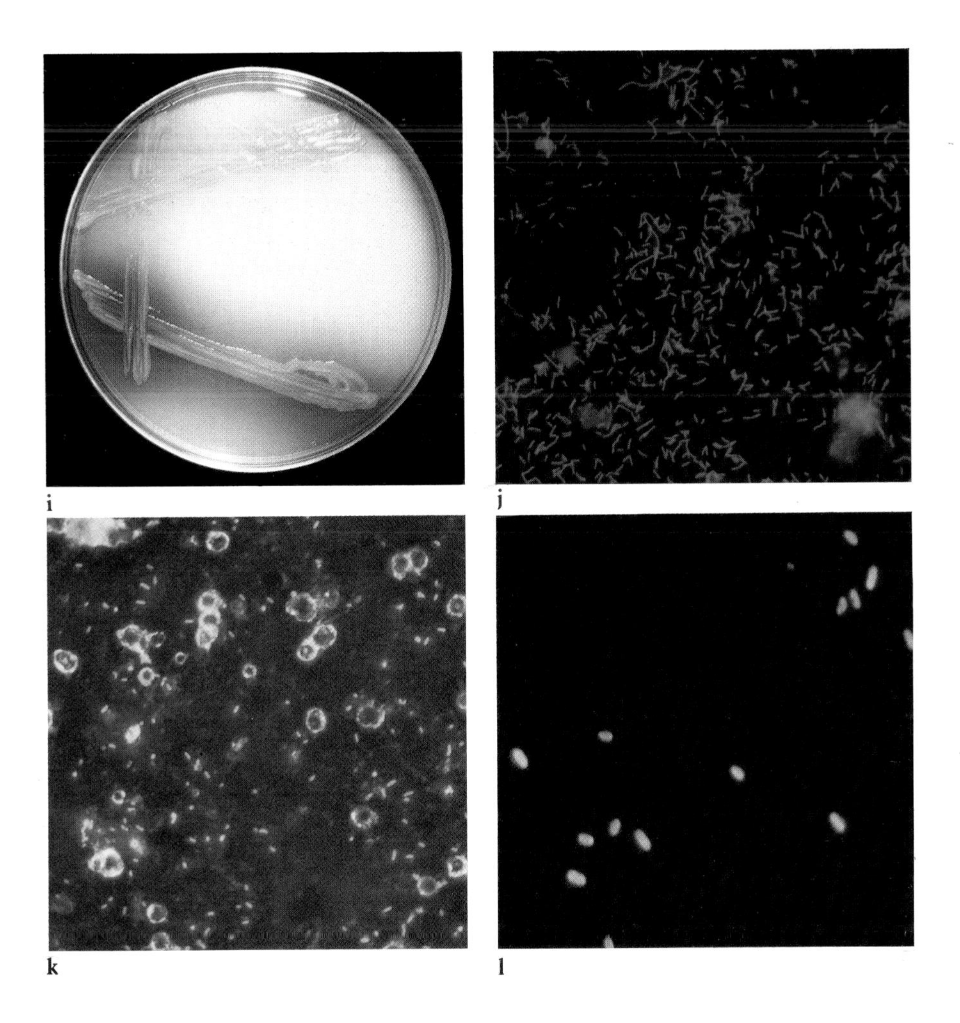

Plate III

i. Browning of TBYE agar medium. Appearance of *L.pneumophila* after 72 hours' incubation.

j. *L.pneumophila* stained by the IFT. Note the presence of both cocco-bacillary and filamentous forms (orginal magnification: ×630).

k. A smear of post-mortem lung tissue stained by the IFT. Note the presence of brightly staining masses of intracellular organisms in addition to numerous extracellular organisms (original magnification: ×630).

l. A smear of bronchial aspirate fluid stained by the IFT (original magnification: ×1000).

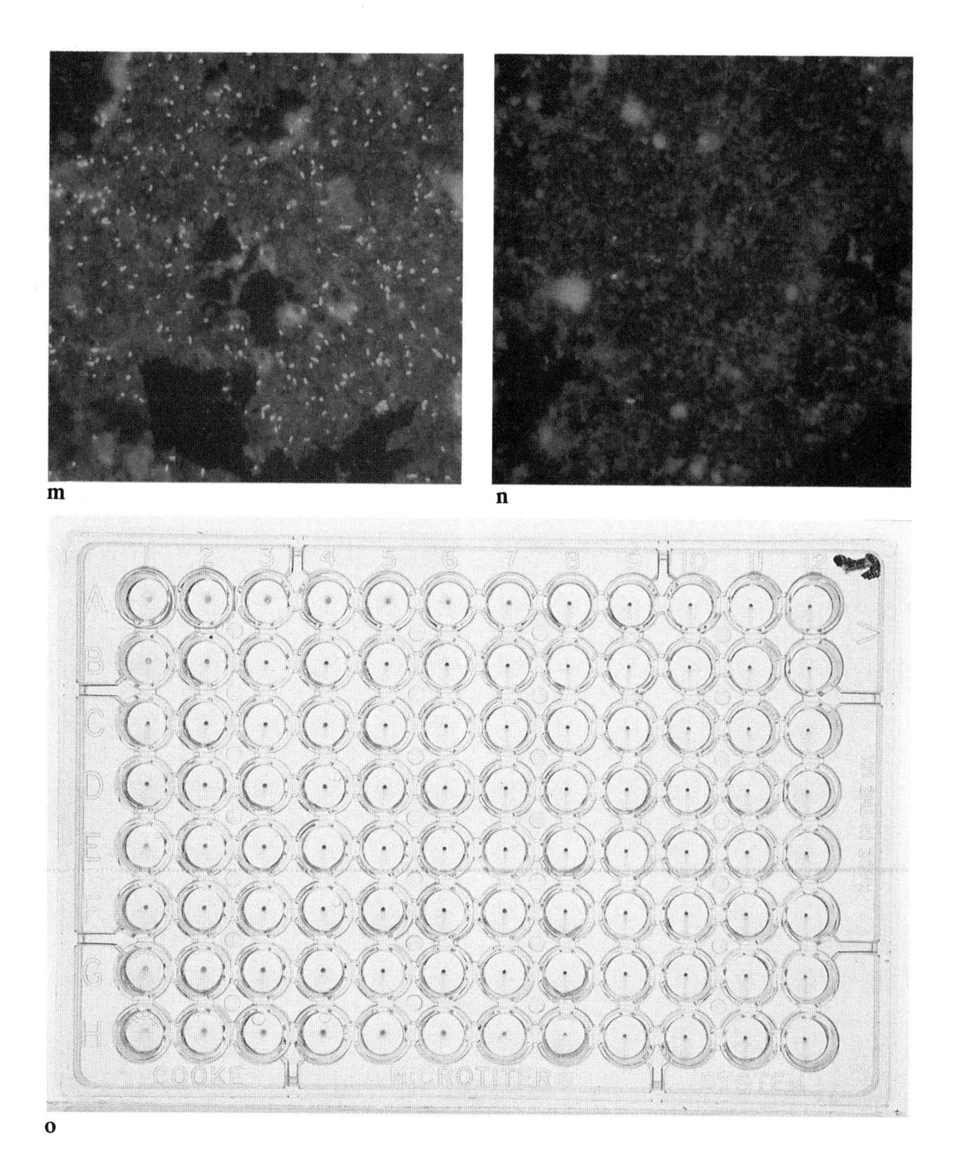

Plate IV

m, n. The indirect immunofluorescent antibody assay (IFAT): formolized yolk-sac antigen (FYSA) reacted with (m) the human reference positive serum, and (n) a negative control serum (original magnification: ×630).

o. Reading the rapid microagglutination test (RMAT) – Row A is the human reference positive serum (titre 128), wells 1–7 show a positive reaction, wells 8–12 a negative reaction. Row B is a negative serum (titre 2).

D Gelatin liquefaction

All *Legionella* species with the exception of *L.micdadei*, *L.feeleii*, and in this laboratory, *L.maceachernii* produce a gelatinase. This can be simply demonstrated by inoculating BCYE medium in which the agar has been replaced by 3% gelatin. If a gelatinase is produced by the test organism the gelatin will be liquefied preventing the broth from solidifying when it is refrigerated.

BCYE – gelatin broths should be pre-incubated at 37 °C for at least 24 hours before use to ensure they are sterile. They are then heavily inoculated with growth from a 48 – 72 hour BCYE plate, mixed well and incubated for up to five days at 37 °C. Known positive (e.g. *L.pneumophila* Knoxville-1 NCTC 11286) and negative (*E.coli*) controls, together with an uninoculated broth, should be processed on each occasion.

The broths are removed each day and refrigerated at 4 °C for one hour before being examined. A positive result is recorded if the test broth remains liquid when both the uninoculated and inoculated negative control broths have solidified. Test broths which solidify are reincubated and examined again after a further 24 hours. If after five days a broth, which is visibly turbid, has not liquefied a negative result is recorded.

E Bromocresol-purple spot test

This test, which was first described by Garrity and colleagues (1980), is used to distinguish *L.micdadei* from other *Legionella* spp.

A 1% stock solution of bromocresol-purple (BCP) is prepared by adding 0.1 g to 10 ml of 1 N KOH. This is diluted with distilled water to 0.04% BCP for use. One drop of the BCP solution is applied to a piece of Whatman No. 1 filter paper. Using a wooden stick growth from a 48 – 72 hour BCYE plate is rubbed into the BCP-saturated filter paper. The colour observed after 30 seconds is recorded; yellow to dull-green is considered negative, and 'aquamarine' blue is positive. In addition to *L.micdadei* the type strain of *L.maceachernii* (NCTC 11982) gives a positive reaction in this test.

IV Other phenotypic characteristics

The tests listed below are included here for completeness. Although widely reported in the literature they have not proved particularly helpful in the identification or speciation of legionellae in our laboratory.

A Oxidase test

The result of the oxidase test obtained with *Legionella* spp. can vary considerably.

These variations are dependent, in part, on: the age of the culture examined, whether the dimethyl or tetramethyl compound is used, and the time allowed for a positive result to develop.

A 1% (w/v) solution of tetramethyl-*p*-phenylenediamine di-hydrochloride is made in distilled water. The solution is prepared immediately before use as the tetramethyl compound is unstable. Using a wooden stick a small amount of growth is removed from a 48–72 hour BCYE agar plate and rubbed on to a filter paper impregnated with fresh oxidase reagent. The production of a blue colour within 10 seconds is a positive result. Known positive (e.g. *Pseudomonas aeruginosa*) and known negative (e.g. *E.coli*) controls should be tested on each occasion.

Most isolates of *L.pneumophila* examined in this laboratory are negative in this test. In some instances a blue coloration has started to appear within the 10 seconds, these isolates are regarded as weakly positive. The test results for the few isolates examined of species other than *L.pneumophila* have usually reacted as reported in the literature and are shown in Table 5.2.

B β-Lactamase production

Most legionellae produce β-lactamase and this can be easily demonstrated using a chromogenic cephalosporin (Nitrocefin).

Reagents

A 0.05% solution of Nitrocefin (Glaxo Ltd) is made by dissolving 10 mg in 1 ml of dimethyl sulphoxide and adding 19 ml of PBS pH 7.2 to this. The reagent can be stored at 4 °C for several months.

Procedure

A loopful of growth from a 48–72 hour BCYE agar plate is emulsified in 50 μl of chromogenic cephalosporin solution in the well of a microtitre plate. The microtitre plate is sealed, vortexed, and incubated at room temperature. After 60 minutes the wells are examined for colour change. A red or orange colour is positive, while a yellow colour is negative. Known positive (*L.pneumophila*) and negative (*L.micdadei*) controls together with uninoculated reagent should be included with each test.

Isolates of *L.micdadei* and *L.maceachernii* are reported not to produce β-lactamase. Test results are variable for isolates of *L.bozemanii*, *L.longbeachae* and *L.feeleii*. Strains of all other species are positive.

Table 5.2 Summary of the biochemical reactions and phenotypic characteristics of the Legionellaceae.

		L. pneumophila	*L. anisa*	*L. bozemanii*	*L. cherrii*	*L. dumoffii*	*L. gormanii*	*L. parisiensis*	*L. steigerwaltii*	*L. micdadei*	*L. maceachernii*	*L. wadsworthii*	*L. feeleii*	*L. oakridgensis*	*L. spiritensis*	*L. erythra*	*L. rubrilucens*	*L. hackeliae*	*L. jamestowniensis*	*L. jordanis*	*L. longbeachae*	*L. sainthelensi*	*L. santicrusis*	*L. israelensis*	'*L. londoniensis*'	'*L. geestiae*'	'*L. quateriensis*'	'*L. nautarum*'	'*L. worsliensis*'
Growth on	BA	−[1]	−	−	−	−	−	−	−	−	−	−	−	−	−	−	−	−	−	−	−	−	−	−	−	−	−	−	−
	BCYE	+	+	+	+	+	+	+	+	+	+	+	+	+	+	+	+	+	+	+	+	+	+	+	+	+	+	+	+
	BCYE−Cys	−	−	−	−	−	−	−	−	−	−	−	−	+	±	−	−	−	−	−	−	−	−	−	−	−	−	−	−
Catalase reaction		+	+	+	+	+	+	+	+	+	+	+	+	+	+	+	+	+	+	+	+	+	+	+	+	+	+	+	+
Nitrate reduction		−	−	−	−	−	−	−	−	−	−	−	−	−	−	−	−	−	−	−	−	−	−	−	−	−	−	−	−
Flagella		+	+	+	+	+	+	+	+	+	+	+	+	−	+	+	+	+	+	+	+	+	+	+	ND[2]	ND	+	ND	ND
Hippurate hydrolysis		+	−	−	−	−	−	−	−	−	−	−	−	−	±	−	−	−	−	−	−	−	−	−	−	−	−	−	−
Autofluorescence[3]		−	W	W	W	W	W	W	W	−	−	−	−	−	−	R	R	−	−	−	−	−	−	−	−	−	−	−	−
Gelatinase production		+	+	+	+	+	+	+	+	−	−	+	−	+	+	+	+	+	+	+	+	+	+	+	+	+	+	+	ND
Browning on TBYE		+	+	+	+	+	−	+	NG	−	−	−	+	+	+	+	+	+	+	+	+	+	+	+	+	+	+	+	ND
Bromocresol-purple		−	−	−	−	−	−	−	−	+	±	−	−	−	−	−	−	−	−	−	−	−	−	−	−	−	−	−	ND
β-Lactamase production		+	+	V	+	+	+	+	+	−	−	+	V	+	+	+	+	+	+	+	V	+	+	+	+	+	+	+	ND
Oxidase		V	+	V	−	−	−	+	−	+	+	−	V	−	+	V	−	V	−	+	+	+	−	−	−	−	−	−	ND

1. Results recorded: + = positive, ± = weakly positive, − = negative, V = variable, NG = No growth on this medium.
2. ND = no data available.
3. Autofluorescence: − = no fluorescence, W = blue/white, R = red.

C Urease

Legionella spp. are reported to be urease negative. This is usually demonstrated by heavily inoculating a Christensen's urea agar slant with growth from a 48–72 hour BCYE agar plate (Orrison *et al.*, 1983b). The obvious limitation of this test is that the urea slants do not support the growth of legionellae therefore only a strong urease reaction would be revealed.

A Laboratory Manual for *Legionella*
Edited by T. G. Harrison and A. G. Taylor

CHAPTER 6

Identification of Legionellae by Serological Methods

T. G. Harrison and A. G. Taylor

I Introduction

As has been illustrated by the previous chapter the speciation of members of the family Legionellaceae is difficult. The use of biochemical reactions, GLC and isoprenoid quinone profiles can confirm that an isolate is a *Legionella* and indicate to which species it belongs, but DNA hybridization studies are needed for a definitive identification. As this facility is not available in many laboratories presumptive identification is usually made using serological methods.

Legionellae can be subdivided into serogroups by their reactions with hyperimmune rabbit antisera containing antibodies directed against the somatic lipopolysaccharide (LPS), or 'O', antigens. At the time of writing there are 51 serogroups comprising 34 species (Brenner, 1986; Appendix 1) and it is certain that others will be described in the future. As many of the species and serogroups have antigens in common, cross-reactions are seen within the genus and the serological identification of legionellae is therefore not always simple.

Cross-reacting antibodies can usually be removed from antisera by absorption with suspensions of bacteria of heterologous strains to render the antisera species- or serogroup-specific. However, there are significant problems using this method. Firstly, absorption is labour intensive for the reagent producer. Secondly, the frequent description of new serogroups and species, many of which are defined with difficulty, results in the need for constant re-evaluation and often additional absorption procedures. Finally, the antisera are evaluated against very few representative strains of each species but antigenic variation within a serogroup may be extensive. In consequence absorbed antisera which appear satisfactory when prepared, may have such a restricted range of specificities that strains subsequently isolated do not react with them. This may result in isolates being misidentified.

The indirect immunofluorescent test (IFT) described below is a simple and reliable method of identifying most legionella isolates. Although some species and serogroups cannot be distinguished using this method alone, these comprise only a small percentage of isolates routinely encountered. The majority of isolates, both clinical and environmental, are *L. pneumophila* and most of these serogroup 1 (Bartlett *et al.*, 1983; Desplaces *et al.*, 1984). This species is serologically distinct and so the appropriate specific reagents can be produced without the necessity for absorption procedures.

II Preparation of hyperimmune rabbit antisera

A Preparation of antigens for inoculation

The strain of *Legionella* against which the antiserum is to be prepared is inoculated on to BCYE agar and incubated at 37 °C for 72 hours. The bacterial growth is gently removed from the surface of the agar and a dense suspension made in 2% formalin in PBS (Dulbecco A, pH 7.2). After overnight incubation (37 °C) to kill the organisms, the sterility of the suspension is checked by inoculating a BCYE agar plate which is incubated for 72 hours at 37 °C. The bacterial suspension is then diluted so that the final concentration of formalin is not greater than 0.25% (higher concentrations are not suitable for intravenous injection into rabbits) and the number of organisms is approximately 3×10^9/ml

(OD_{605} = 1.2). It is not generally necessary to incorporate an adjuvant into the inoculum as *Legionella* LPS is itself immunostimulatory.

Formalin-killed whole bacteria are used as inoculum and these will frequently be flagellate. In consequence the antisera produced will often have high anti-flagella ('H') titres. All *Legionella* spp. share common 'H' antigens and the antisera will therefore react with any flagellate legionellae irrespective of serogroup. In the IFT this causes no problem as flagellar fluorescence is easily distinguished from somatic fluorescence. However, if the sera are to be used for agglutination tests problems may be encountered. In these cases a non-flagellate mutant can be selected by multiple passage of an isolate on BCYE agar. Alternatively the flagella can be removed by acid washing or heating the inoculum (see below).

B Animals used

Antisera are raised in New Zealand White rabbits of weight 3–4 kg. Inoculations are administered intravenously (i.v.) and blood samples (pre- and post-inoculation) are taken from the marginal ear vein. Five days after the final inoculation the rabbits are anaesthetized (sodium pentabarbitone) and exsanguinated by cardiac puncture.

C Inoculation schedule

Considerable variation is seen in the antibody response of individual rabbits. It is therefore advisable to use at least three rabbits for each antiserum to be made. The inocula are administered in two courses with a rest of one month between the first and second injections.

1st course	*2nd course*
Day 1 : 0.1 ml	Day 40 : 0.5 ml
Day 3 : 0.2 ml	Day 43 : 1.0 ml
Day 5 : 0.5 ml	Day 45 : 1.5 ml
Day 8 : 1.0 ml	Day 47 : 1.5 ml
Day 10 : 1.5 ml	
Day 12 : 1.5 ml	

A blood sample is taken from each rabbit on day 17 and the sera tested by IFT to check that antibody is present. Typically the titres of these test bleeds are greater than or equal to 1024. After the second course of inoculations the blood is collected from each rabbit and allowed to clot. The serum is then removed and clarified by centrifugation. Provided the serum IFT titres are similar from each rabbit, the sera are pooled, filter sterilized and stored in 50 ml volumes at −40 °C until required for use.

Although the above schedule has given satisfactory results for most antigens, some strains of *Legionella* do not elicit a strong antibody response. McKinney (1985) has reported similar experiences and suggests that Freund's complete adjuvant be used in these circumstances. This would seem acceptable if the antiserum is to be used only to confirm the identity of isolates. However, if the antiserum is also to be used for the diagnosis of clinical infections the use of adjuvant may not be desirable as in this situation the specificity of the reagent is of paramount importance. The presence of antibodies to *Mycobacterium tuberculosis* used as the adjuvant is undesirable in a reagent which may be used to examine clinical specimens which could contain tubercle bacilli.

III Test procedures

A Preparation of antigen suspensions

A small quantity of bacterial growth is removed from a BCYE plate incubated for 48−72 hours, and emulsified in 1 ml of 2% formalin in PBS. After overnight incubation to kill the organisms a small quantity of this stock suspension is taken and diluted to an optical density $OD_{605} = 0.2$. This suspension is then further diluted 1:10. Alternatively a small quantity of stock suspension is diluted until it is only just visibly turbid. Normal yolk-sac (NYS) (0.1%) is then added to provide a matrix which will attach the bacterial cells to the microscope slide. Gelatin can be used as an alternative to NYS, however we prefer NYS because when examining the slides it provides a dull-green background which aids focusing.

B Preparation of slides for immunofluorescence

PTFE-coated microscope slides with 3 mm wells are used. If slides with alternative size wells are used the volumes of reagents should be altered accordingly. To each well is added 5 μl of antigen suspension. The slides are air-dried at 37 °C (20 minutes) and fixed in acetone at room temperature (15 minutes).

Serum dilutions

Dilutions of rabbit anti-*Legionella* antisera are made in PBS as specified by the reagent producer (typically 1:200), and 10 μl is added to each appropriate well. The slides are then incubated in a moist chamber at 37 °C for 20 minutes.

Washing

The slides are rinsed with PBS and then washed for 15 minutes with two changes of PBS. They are then briefly rinsed with distilled water, gently blotted and dried at 37 °C (10 minutes).

FITC conjugate

To each antigen-coated well is added 5 μl of anti-rabbit-Ig conjugate dilution, the slides are then incubated in a moist chamber at 37 °C for 20 minutes and washed as above. After drying the slides should be examined immediately or stored in the dark until they can be viewed which should be within 24 hours of preparation.

Microscopy

The slides are mounted in glycerol mounting media or viewed unmounted with a water immersion objective. They should be examined by epi-illumination using ×10 eyepieces and a ×100 oil objective or a ×63 water objective. We use a Zeiss microscope equipped with a HB050 mercury vapour lamp and the following interference filters: excitation filters – LP 455 nm and KP 490 nm, dichromic mirror – FT 510 nm, barrier filters – BP 520–560 nm and BP 590 nm.

Preparation of glycerol mounting media:

Buffered saline* pH 8.5	1 part
Glycerol neutral	9 parts

Prepare 0.067 M K_2HPO_4 by dissolving 1.61 g in 100 ml 0.85% NaCl.
Prepare 0.067 M KH_2PO_4 by dissolving 0.907 g in 100 ml 0.85% NaCl.

If water objectives are used care must be taken to ensure that the pH of the water is above pH 8.5.

C Conjugate titration and storage

An appropriate, fluorescein-conjugated anti-rabbit immunoglobulin preparation should be titrated in PBS against a known legionella and antiserum to determine its optimal working dilution. *Legionella*-coated slides are prepared as described above. To these are added dilutions of the homologous rabbit antiserum, e.g. 1:50–1:3200. The highest dilution of conjugate at which the maximum titre of the antiserum is obtained is considered optimal. This will vary from batch to batch but is typically between 1:20 and 1:80.

After reconstitution the conjugate is divided into volumes of about 50 μl (enough to make 1–2 ml working strength) and stored at –40 °C until required. Preservative (e.g. 0.08% NaN_3) should be added to working dilutions of the conjugate which can then be stored at 4 °C for up to 14 days.

* Ten parts 0.067 M K_2HPO_4 in 0.85% NaCl plus one part 0.067 M KH_2PO_4 in 0.85% NaCl.

IV Limitations of the technique

DMRQC provide a range of hyperimmune rabbit antisera for serogrouping legionellae. The reactions of these reagents with isolates of bacteria from genera other than *Legionella* are largely unknown and are not investigated in the evaluation of these reagents. It is assumed that any isolate being examined has already been shown to be a presumptive legionellae (see Chapter 5).

Although the identification of legionellae by immunofluorescence is essentially qualitative, and therefore not greatly affected by the subjective nature of the technique, there are instances where the method must be used semi-quantitatively for accurate identification of an isolate. For this reason the protocol detailed above must be followed carefully and attention paid to the quality of the observed fluorescence. Plate III.j shows the typical appearance of legionellae stained with their homologous antiserum. The organisms are evenly stained with sharply delineated edges. Where bacteria stain unevenly giving a mottled appearance, or are ill-defined and diffusely stained this indicates partial homology and they can be expected to react more typically with other antisera.

Isolates frequently react with only one antiserum at its working dilution (1:200) and hence identification is easily established. In other instances the isolate will react with varying degrees of intensity with several antisera. In these cases dilutions are made with each of the reacting antisera and the isolate reexamined. If the isolate reacts at high serum dilutions with only one antiserum its identity can then be deduced. Where an isolate reacts with several antisera at high dilutions it may be tentatively identified by its pattern of cross-reactivity (Table 6.1). However, if it is important to identify an isolate precisely it should be sent to a reference laboratory for further studies.

V Interpretation of results

The following discussion relates to the interpretation of results obtained using DMRQC reagents. Where reagents are obtained from other sources different cross-reactions may be encountered and the information given here may not be applicable. The cross-reactions encountered using current DMRQC reagents are shown in Table 6.1 and most of these have been reported previously by workers using antisera prepared elsewhere. Where cross-reactions shown here have not been documented before they are usually minor. These reactions may have gone unnoticed because most reports relate to CDC-prepared direct conjugates or sera used for slide agglutination assays. The IFT technique detailed here is probably more sensitive than either of these methods and so may be expected to recognize weaker cross-reactions.

The following notes are intended to highlight areas where confusion has arisen, or may arise, in the serogrouping of legionellae. It should be stressed that with the exception of *L.pneumophila*, the number of strains of each serogroup examined is small, and in some instances only a single isolate has been examined.

L.pneumophila serogroup 1 The majority of isolates obtained from either clinical or environmental sources are likely to be *L.pneumophila* serogroup 1. The antigenic diversity of this serogroup is such that some isolates may only react weakly with the available serogroup 1 antisera. However, as these isolates do not react with any other antisera identification is usually simple. The DMRQC antiserum reacts well with most clinical isolates.

Occasionally a serogroup 1 strain will react weakly with *L.pneumophila* serogroup 9 antiserum and conversely some serogroup 9 strains react with serogroup 1 antiserum. In the original description of *L.pneumophila* serogroup 9 (Edelstein *et al.*, 1984a) the type strain was found to be highly reactive with antiserum raised against the serogroup 1 strain 'OLDA'.

L.pneumophila serogroup 5 These isolates usually react only with the serogroup 5 antiserum. However, the serogroup 5 strain used by many workers to raise antisera is the strain Dallas-1E. This strain has recently been shown to be the subject of some taxonomic confusion and is probably a representative of a *L.pneumophila* subspecies or possibly a new *Legionella* species (Selander *et al.*, 1985). Antisera raised against the Dallas-1E strain cross-react with serogroup 8 isolates (see below), but antisera raised against the serogroup 5 strain Cambridge-2 (NCTC 11417) appear not to. Cambridge-2, which was the first reported serogroup 5 strain (Nagington *et al.*, 1979), is used by DMRQC for reagent production and the resulting antiserum is serogroup specific.

L.pneumophila serogroup 6 Isolates from this serogroup always react with antisera raised against both *L.pneumophila* serogroup 3 and *L.pneumophila* serogroup 6. The first isolate of this serogroup (Tobin *et al.*, 1980) was shown to be a serogroup 3 isolate which possessed additional distinct antigens (Taylor & Harrison, 1979). This was subsequently confirmed by workers at the CDC but it was considered sufficiently distinct to be designated a new serogroup (McKinney *et al.*, 1980). Isolates of serogroup 6 are easily distinguished from those of serogroup 3 as the latter react only with their homologous antiserum.

L.pneumophila serogroup 8 References are made in the literature to *L.pneumophila* serogroup '4/5 cross' isolates (Dournon *et al.*, 1983). These isolates have since been classified as *L.pneumophila* serogroup 8 strains. The cross-reactions with serogroup 5 antisera are dependent on the serogroup 5 strain used to raise the antisera (see above). In addition, isolates of this serogroup may react with antisera against serogroup 4, serogroup 10 and occasionally serogroup 2.

The blue/white autofluorescent species Some of these *Legionella* species are serologically indistinct. For example the following isolates cannot be easily distinguished: *L.anisa*, *L.bozemanii*, serogroup 1 and serogroup 2, and *L.parisiensis*. Comparison of titres obtained with antisera against each of these may give an indication of the species, but final identification will require other techniques.

Table 6.1 Cross-reactions between serogroups of *Legionella* seen using hyperimmune rabbit antisera in the IFT.

Antiserum	Species or serogroups with which heterologous reactions may be seen
L.pneumophila Sgp 1	(*L.pneumophila* Sgp9)
2	(*L.pneumophila* Sgp4,8,10)
3	*L.pneumophila* Sgp6*
4	(*L.pneumophila* Sgp2,8)*
5	
6	
7	
8	(*L.pneumophila* Sgp2,4)*
9	(*L.pneumophila* Sgp1)*
10	*L.pneumophila* Sgp4,8
11	
L.anisa	*L.bozemanii* Sgp1,2* (*L.jordanis*)* *L.longbeachae* Sgp2 (*L.micdadei*) *L.parisiensis**
L.bozemanii Sgp 1	*L.anisa** (*L.bozemanii* Sgp2)* *L.jordanis** *L.parisiensis**
2	*L.anisa** *L.bozemanii* Sgp1* *L.longbeachae* Sgp2* *L.parisiensis**
L.jordanis	*L.anisa** *L.bozemanii* Sgp1,2* (*L.longbeachae* Sgp1,2) *L.parisiensis** (*L.sainthelensi*) (*L.steigerwaltii*) (*L.wadsworthii*)
L.longbeachae Sgp1	(*L.longbeachae* Sgp2) (*L.sainthelensi*)* *L.santicrucis**
2	*L.anisa** *L.bozemanii* Sgp2* (*L.longbeachae* Sgp1)* (*L.sainthelensi*)* *L.santicrucis*
L.micdadei[1]	
L.oakridgensis	*L.sainthelensi**
L.parisiensis	*L.anisa** *L.bozemanii* Sgp1,2* *L.jordanis**

L.cherrii		*L.steigerwaltii**	*L.rubrilucens*	
L.dumoffii			*L.sainthelensi*	*L.longbeachae* Sgp1,2*
				L.oakridgensis
L.erythra				(*L.santicrucis*)*
L.feeleii	Sgp1	(*L.feeleii* Sgp2)*	*L.santicrucis*	(*L.bozemanii* Sgp1)
	2	(*L.feeleii* Sgp1)*		(*L.jordanis*)
				L.longbeachae Sgp1,2*
				*L.sainthelensi**
L.gormanii				
L.hackeliae	Sgp1	(*L.dumoffii*)	*L.spiritensis*	
		(*L.hackeliae* Sgp2)*		
	2	(*L.hackeliae* Sgp1)*	*L.steigerwaltii*	*L.cherrii**
		(*L.jordanis*)		
			L.wadsworthii	(*L.bozemanii* Sgp2)
L.israelensis[2]				(*L.longbeachae* Sgp1,2)
				(*L.santicrucis*)
L.jamestowniensis		'*L.geestiae*'		(*L.sainthelensi*)

L.anisa = Major cross-reactions; antiserum reacts with this species/serogroup at a titre >800.
(*L.anisa*) = Minor cross-reactions; antiserum reacts with this species/serogroup at a titre 200–400.
1. This species is reported to cross-react with *L.maceachernii* (Brenner *et al.*, 1985).
2. This species is reported to cross-react with *L.wadsworthii* (Bercovier *et al.*, 1986).
* These cross-reactions have been previously reported (Bisset *et al.*, 1983; Campbell *et al.*, 1984; Tang *et al.*, 1984; Thacker *et al.*, 1985a,b; Wilkinson *et al.*, 1985).

L.cherrii and *L.steigerwaltii* are serologically distinct from the above three species but cannot be distinguished from each other by serological means. The other blue/white species, *L.gormanii*, does not cross-react with any of the above species.

'*L.geestiae*' The type strain cross-reacts with antisera raised against *L.jamestowniensis*. The serological relationship between these species has yet to be investigated in detail.

L.jordanis The type strain reacts strongly with antisera raised against *L.bozemanii* serogroup 1 and *L.parisiensis*. However, as *L.jordanis* is not a blue/white autofluorescent species it can be distinguished.

L.longbeachae serogroup 1 and serogroup 2, *L.sainthelensi* and *L.santicrucis* are antigenically similar. Antisera raised against any of these are likely to react with isolates of the other species but comparisons of the titres obtained might give a good indication of an isolate's identity. In addition to the above reactions *L.longbeachae* serogroup 2 shares antigens in common with most of the blue/white autofluorescent species. *L.sainthelensi* and *L.oakridgensis* also cross-react strongly with each other.

Antisera against '*L.geestiae*', '*L.londoniensis*', '*L.nautarum*', '*L.quateriensis*' and '*L.worsliensis*' are not available at the present time.

VI Preparation and use of anti-flagellar antisera

It has previously been reported that some species of Legionellaceae share common flagellar antigens (Rodgers & Laverick, 1984). We have extended these observations using both ELISA and IFT techniques (Harrison, unpublished data). Of the 23 species examined representatives of 16 have been shown to possess cross-reacting 'H' antigens (Table 6.2). It is probable that all species (with the possible exception of *L.oakridgensis*) share 'H' antigens as the representatives of the other six species examined using silver-plating stains appeared to be aflagellate. In addition *L.longbeachae* has previously been shown to share flagellar antigens (Rodgers & Laverick, 1984) although the isolate examined in this laboratory is non-flagellate.

In this laboratory antiserum raised against a highly flagellate strain of *L.pneumophila* has been used to confirm the identity of putative legionellae which we were unable to serogroup with the existing antisera. These isolates which gave positive flagella fluorescence were subsequently found to be either *L.pneumophila* serogroup 10 isolates or a previously unrecognized species.

As discussed above, antisera raised against whole formalin-killed legionellae usually have high anti-'H' titres and these sera can be used to visualize flagella. The only limitation with using such sera is that if the isolate being examined shares somatic antigens with the strain used to raise the antiserum, any flagellar fluorescence will be masked by the more intense somatic fluorescence. Although this may not preclude identification of the isolate it may be desirable to confirm the presence of flagella. In such instances 'H'-specific antisera can be used. These can be prepared by absorbing the antisera with bacteria of the homologous somatic serogroup from which the flagella have been removed by heat treatment.

Table 6.2 Common flagellar antigens within the family Legionellaceae.

Species	Presence of flagella demonstrated by: silver stain[1]	IFT[2]
L.pneumophila (serogroup 1–11)	+	+
L.anisa	+	+
L.bozemanii (serogroup 1–2)	+	+
L.cherrii	+	+
L.dumoffii	+	+
L.erythra	+	+
L.feeleii (serogroup 1)[3]	+	+
L.gormanii	+	+
L.hackeliae	+	+
L.israelensis	−	−
L.jamestowniensis	+	+
L.jordanis	+	+
L.longbeachae (serogroup 1–2)[4]	−	−
L.maceachernii	−	−
L.micdadei	+	+
L.oakridgensis[5]	−	−
L.parisiensis	+	+
L.rubrilucens	+	+
L.sainthelensi	−	−
L.santicrucis	−	−
L.spiritensis	+	+
L.steigerwaltii	−	−
L.wadsworthii	+	+

\+ = Flagella present; − = Flagella absent.
1. Flagella demonstrated using the silver stain as described in Chapter 4.
2. Flagella demonstrated by IFT using rabbit anti-*L.pneumophila* 'H' antiserum.
3. The single strain of *L.feeleii* serogroup 2 examined was found to be non-flagellate by both methods.
4. *L.longbeachae* serogroup 2 previously reported to share H antigens (Rodgers & Laverick, 1984).
5. *L.oakridgensis* is thought to be a non-flagellate species.

The bacterial growth from three heavily inoculated BCYE agar plates is harvested into 10 ml of distilled water. The bacteria are killed and flagella removed by heating at 100 °C for 15 minutes. The bacteria are pelleted by centrifugation, the supernatant is discarded and the bacteria resuspended in distilled water. This washing procedure is repeated twice more. A 1:1 mixture of bacteria and antiserum is then incubated at 37 °C for 2 hours, the bacteria are pelleted by centrifugation and the absorbed antiserum finally clarified by membrane filtration. The antiserum is then tested in the IFT using the homologous strain as the antigen to confirm that somatic antibodies are not detected at working dilutions appropriate for the detection of flagella.

An alternative method for producing 'H'-specific antisera which has also been successfully used in this laboratory is to absorb out 'O' antibodies with bacteria from which the flagella have been removed by acid washing (Fey & Suter, 1979). The authors of this technique used it to prepare purified flagellin from *Salmonella* to raise high titred 'H' antisera. We have not found this to give very satisfactory results in the case of *Legionella* as the flagellin produced is contaminated with LPS and so the antisera produced still has to be absorbed to render it 'H' specific.

The test procedures and conditions used to stain flagella are as described above for the IFT. Although not yet extensively evaluated, this procedure appears to be a specific method to identify isolates of all *Legionella* species.

VII Conclusions

It might incorrectly be concluded from the above discussion that the routine serogrouping of legionellae is complex and often unsuccessful. However, although many cross-reactions are found within the genus, in practice these are not often encountered. Some of the cross-reacting species described have only been isolated from one locality and most of the others are only rarely isolated.

Almost all the isolates submitted to this laboratory for routine examination have been *L.pneumophila*. It is important to serogroup isolates of this species as there is good evidence that some serogroups (particularly serogroup 1) are more likely to give rise to infection than are others.

A Laboratory Manual for *Legionella*
Edited by T. G. Harrison and A. G. Taylor

CHAPTER 7

Confirmation of the Identity of Legionellae by Whole Cell Fatty-Acid and Isoprenoid Quinone Profiles

R. Wait

I Introduction

Legionellae are presumptively identified by their failure to grow in artificial culture except on media containing iron and cysteine. Definitive identification requires phenotypic and serological techniques, which can be complex because of the large number of species and serogroups; moreover new species and new serogroups of known species are still emerging, rendering reliance on immunological methods inadequate. The analysis of cellular fatty acids by gas chromatography is thus a valuable additional approach to the identification of these organisms.

L.pneumophila is characterized by a pattern of cellular fatty acids in which, unusually for a Gram-negative organism, branched chain acids predominate (Moss *et al.*, 1977), the most abundant being 14-methyl pentadecanoic acid (i16:0). The profiles of other *Legionella* species are qualitatively similar, but exhibit quantitative differences which in some cases allow discrimination between them. The structures of the major acids occurring in *Legionella* species are listed in Figure 7.1 together with their conventional abbreviations.

A further interesting feature of the Legionellaceae is the presence of some unusual members of the ubiquinone series, analysis of which can further aid their identification. A brief account of the techniques for measuring these compounds therefore is also provided.

II Gas–liquid chromatography of fatty acid methyl esters

A Gas–liquid chromatography

Gas–liquid chromatography (GLC) is a simple and powerful method for the separation and quantitation of mixtures of volatile organic compounds. The components of the mixture are separated by partitioning between a mobile carrier gas, and a stationary liquid phase. In packed column chromatography the liquid phase coats an inert granular support, whereas in capillary chromatography it is deposited as a thin film on the column wall. As a compound moves through the column, it will spend a proportion of the time as a vapour in the carrier gas, and the rest dissolved in the liquid phase (when it will not be moving at all). Hence the greater its affinity for the liquid phase the more slowly it will move through the column. Different compounds are thus separated because of (i) differences in their vapour pressure at the temperature of the analysis (which determine their concentration in the gas phase) and (ii) differences in their affinity for the liquid phase (which are a complex function of the chemical structures of both compound and stationary phase).

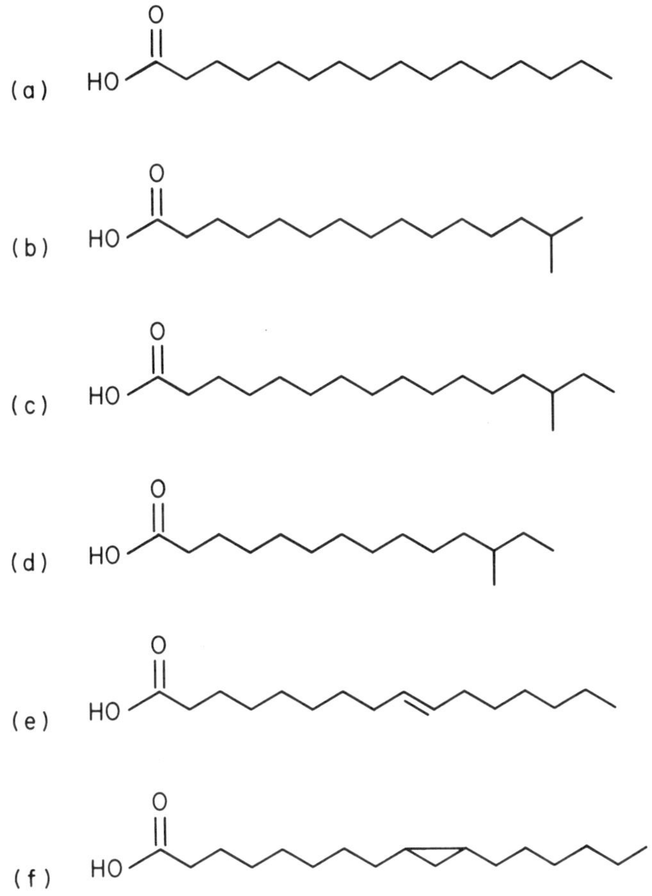

(a) HO

(b) HO

(c) HO

(d) HO

(e) HO

(f) HO

Figure 7.1 Structures of bacterial fatty acids
(a) hexadecanoic acid (n16:0)
(b) 14-methyl pentadecanoic acid (iso-hexadecanoic acid: i16:0)
(c) 14-methyl hexadecenoic acid (ante-iso-heptadecanoic acid: a17:0)
(d) 12-methyl tetradecanoic acid (ante-iso-pentadecanoic acid: a15:0)
(e) 9-hexadecenoic acid (n16:1)
(f) 9,10 methylene hexadecanoic acid (cyclopropane heptadecanoic acid: cyc 17).

The retention time is the total time between the injection of a solute, and its emergence from the column. For a given stationary phase, and provided that all experimental conditions are unchanged, the retention time will be a characteristic physical property of the compound and can be used for the identification of unknowns. However, it should be remembered that while non-identical retention times provide definitive proof of non-identity of two compounds, different compounds can and do have identical retention times. Supplementary techniques such as mass spectrometry or spectroscopic methods should therefore always be used to confirm identifications made by GLC.

Modern capillary instruments fitted with flame ionization detectors can easily provide nanogram sensitivity, while the use of selective detection (e.g. electron capture or mass spectrometry) allows detection limits in the picogram range to be attained. The ability to handle complex mixtures at this level of sensitivity makes GLC an indispensable tool for the analysis of biological samples. It is not surprising therefore to find that GLC is increasingly being used routinely in the microbiology laboratory.

The essential components of a gas chromatograph are a sample introduction system, a column, which is contained in a heated and thermostatically controlled oven, and a detector which produces an electrical signal proportional to the quantity of each component as it is eluted. Although it is not intended to attempt a comprehensive account of GLC methods a few points relevant to the successful analysis of bacterial fatty acids will be highlighted here. Those desiring more detail should consult the references below.

Microbiological applications of GLC are covered in the books by Drucker (1981) and Odham and colleagues (1984), while more general accounts of the theory and practice are to be found in Grob (1985) and Jennings (1980, 1981).

B Some theoretical considerations

The degree of separation of two components is governed firstly by the difference in their distribution coefficients between the stationary and the mobile phase, and secondly by the efficiency of the column, which is a measure of the degree to which an initially tight band of solute becomes broadened as it migrates down the column; efficiency being inversely related to the degree of band broadening. Column efficiency is often expressed in terms of 'theoretical plates' (a concept adapted from fractional distillation). The number of plates in a column (n) may be obtained from the expression $n = 5.54\,(t_r/W_{0.5})^2$, where t_r is the retention time of a peak and $W_{0.5}$ is its width at half height. The relative efficiencies of columns of different lengths may be more easily compared by use of the 'height equivalent of a theoretical plate' (HETP) defined as HETP $= L/n$, where L is the column length. The smaller the HETP, the more efficient the column. Column efficiency is related to other system variables by means of the van Deemter equation which may be written in simplified form as HETP $= A + B/\bar{u} + C\bar{u}$. Where $\bar{u}$ is the average linear gas velocity, and A, B and C are complex expressions. A, the so called multipath term, relates peak broadening to the paths of different lengths taken by molecules negotiating their way round irregularly shaped particles of packing. The longitudinal diffusion term, B, describes the ease of longitudinal diffusion in the gas phase and so is also related to the regularity of the packing. The resistance to mass transfer term, C, describes the ease of transfer of solute molecules between gas and liquid phase, and is affected by the ratio of the volume of mobile phase to stationary phase (large ratios leading to higher efficiencies) and also to the evenness of the film thickness (uneven films leading to peak broadening).

The greater efficiencies of open tubular columns compared to packed columns are readily explained by the van Deemter equation. The term A can be ignored, as there is no packing to cause unequal path-lengths. Similarly the value of B is decreased because of the elimination of factors related to packing irregularity. Thirdly, the resistance to mass transfer term, C, is also reduced because the greater 'openness' of capillary columns leads to larger ratios of mobile/stationary phase, and because the liquid phase can be applied much more evenly (mainly because there are no points of contact between adjacent particles coated in liquid phase as occurs with packed columns).

The total plate number achievable with a packed column is limited by the difficulty of using columns longer than about 3 metres; beyond this length the packing usually offers too much resistance to gas flow, necessitating impossibly high pressures at the column head. However, capillary columns up to 100 metres long can be used because of the low pressure drop along their length. Finally, the efficiency of packed columns is limited because most packing materials are poor conductors of heat and so temperature gradients will exist across the thickness of the column (particularly in temperature programming). In consequence molecules at different points in the column will experience slightly different chromatographic conditions.

C Column selection

The column, being the site of component separation, is the heart of any gas chromatographic system and much of the skill of chromatography once resided in the selection of the appropriate column for an analysis. The advent of high efficiency capillary columns has greatly simplified this choice, as most analyses can now be accomplished on such a column coated with one of a relatively small number of stationary phases. Early capillary columns were made of glass, and were extremely delicate, requiring considerable care in installation. More recently flexible fused silica columns coated externally with a tough polyimide sheath have been introduced. These are extremely robust and long lasting, and have taken most of the hazard out of column installation and removal. In addition, as they are extremely flexible, the end can be threaded right up into the detector, thus eliminating many problems due to active surfaces.

i Capillary columns

Mixtures of bacterial fatty acid methyl esters can be successfully analysed on most commercially available capillary columns, coated with phases of low to medium polarity (e.g. OV1, SE30, OV17, SE54). In this laboratory we use vitreous silica columns coated with a covalently immobilized and cross-linked methyl silicone phase (BP-1 from SGE Ltd). Although the column has an upper temperature limit of 320 °C, operation above 200 °C is unnecessary, so very low levels of bleed are encountered, together with correspondingly long column lifetimes. Although 25-metre columns provide larger plate numbers 12-metre columns are cheaper and provide baseline resolution of all the fatty-acids (FA) found in legionellae, together with reduced analysis time and consequently increased sample throughput. However, if the chromatograph is not to be used solely for *Legionella* analysis a 25-metre column is probably a better choice, as there will be less likelihood of having to exchange it to perform other separations. Neither column diameter nor thickness of stationary phase is especially critical; we find 0.2 mm diameter with 0.25 μm film thickness provides excellent resolution and sufficient sample capacity. If it is intended to use 'on column' injection, columns of 0.3 mm diameter are probably a wiser choice. So called 'wide bore' columns are also available. These are vitreous silica columns with an internal diameter of 0.53 or 0.75 mm, which are compatible with flow rates of >10 ml min^{-1}, much higher than are usual in capillary chromatography, and provide efficiencies intermediate between packed and capillary columns. Such columns may offer a compromise solution for users desiring to convert older packed column chromatographs to capillary use, as they are compatible with packed column detectors and injectors.

ii Packed columns

Capillary chromatography with vitreous silica columns must be considered the method of choice. However, if a capillary instrument is not available, then quite satisfactory results may be obtained with packed columns (Moss *et al.*, 1977; Fisher-Hoch *et al.*, 1979).

We have used 2 m × 2 mm i.d. silanized glass columns packed with 100/120 mesh supelcoport coated with 3% OV1 or SP2100 DOH (a high quality deactivated methyl silicone phase). When packed columns are used it is strongly recommended that all analyses be repeated on a second column of contrasting polarity, e.g. 5% DEGS.

The temperature at which the column is operated will depend on the particular separation being attempted. The long chain FA of most interest in *Legionella* species (C14–C18) do not have a broad range of boiling points, and so can usually be analysed isothermally. Temperature programming however may sometimes be useful to improve the resolution of the later peaks. Suitable temperature programmes are given in the summary of chromatographic conditions below.

D Sample injection methods

One of the consequences of the low carrier gas flow rates and low sample capacities of capillary columns is that the sample introduction requirements are more stringent than those of packed column chromatography.

The function of any capillary injection system is to deliver the correct amount of sample (typically about 10 ng per component) in the tightest possible band to the top of the column. The three most commonly utilized techniques are split, splitless and on-column injections, all of which have been successfully employed for the analysis of bacterial fatty acids.

i Split injection

The use of an inlet splitter ensures that only a proportion of the sample enters the column while the remainder is vented to the atmosphere. The split ratio, typically between 1:20 and 1:100, is selected by the operator. This allows a much higher flow rate to be maintained through the injector than through the column, so that the sample is rapidly swept on to the column and deposited as a sharp band. Since the sample is flash vaporized in the heated injector there is some danger of loss of thermally labile components, but fortunately none of the major FA in *Legionella* species are especially vulnerable to thermal decomposition. Another possible problem of a split injection is that components of higher boiling point may be discriminated against, so that the components entering the column are not representative of the sample as a whole. Such problems are usually minimized by careful injector design and by operating the injector at large split ratios (1:100 and greater) and at temperatures above the boiling points of all sample components.

When samples are very dilute, for example because only a few bacterial colonies are available for sample preparation, a split injection may provide inadequate sensitivity. In such cases splitless injection is to be preferred.

ii Splitless injection

In a splitless injection the 'solvent effect' (Grob & Grob, 1981) is responsible for focusing the injected sample into a narrow band at the head of the column. The injector is maintained at a temperature above the boiling point of the injection solvent, while the column oven is about 20 °C below this. The solvent condenses in the upper part of the column causing an increase in the film thickness and thus a temporary increase in the affinity of sample molecules for the liquid phase. Unable to move down the column the sample forms a very narrow band at its head. The column is then rapidly heated to its normal operating temperature, the solvent vaporizes and the sample components recover their normal affinity for the liquid phase, and the separation commences as the sample migrates down the column. Because relatively large amounts of solvent (up to 1.5 μl) are introduced into the system it is usual to flush the last vestiges of solvent from the injector by opening the split valve between 30 and 60 seconds after injection, thus reducing the effects of solvent tailing.

It is desirable that the affinity of the sample components for the solvent and the stationary phase should be broadly similar. Thus for a low polarity methyl silicone type phase, a hydrocarbon solvent is suitable. We have found that iso-octane (boiling point 98 °C) is a very satisfactory solvent for splitless injection of fatty acids in the C12–C20 range on to a methyl silicone column.

Splitless injection is a vaporizing injection, so difficulties may be encountered with thermally labile samples. However, as the sample is delivered to the column over a relatively long time period it is not necessary to have the injector at an especially high temperature (200–250 °C is usually sufficient), which tends to reduce such problems and also limits the amount of septum bleed.

Split and splitless injection procedures have been systematically compared by Larsson and Odham (1984) who observed no loss of resolution in the splitless method.

iii On-column injection

The third sample introduction system commonly used in capillary chromatography is the on-column injector, where the sample is injected directly into the column with a fine gauge needle. These needles are too delicate to pierce a rubber septum, so a septumless valve is provided which eliminates septum bleed. Some designs allow the use of highly inert fused silica needles which can prevent metal catalysed decomposition of sample components. Since the sample is introduced quantitatively into the liquid phase, without flash evaporation, this is an injection method which is especially suitable for quantitative analysis of thermally labile components, or samples which span a wide boiling range. Band narrowing is achieved by means of the solvent effect. However, since virtually no mixing with the carrier gas occurs, a solvent effect is produced across a wider range of conditions than in splitless injection (i.e. with low injection volumes and above the boiling point of the solvent).

On-column techniques generally require more stringent pre-chromatographic purification of samples, as any polar or involatile materials present are introduced into the column, and can cause rapid deterioration of its performance.

Capillary inlet systems tend to be intolerant of poor injection technique, especially when inlet splitters are used. The 'plunger in needle' type syringes, in which the sample is contained in the needle, should be avoided as volatile components can be distilled out leading to injections which are unrepresentative of the sample. Reproducible injections can be obtained with the 'hot needle' technique in which the sample is withdrawn into the syringe barrel, the needle is inserted through the septum, allowed to warm up for several seconds, and the injection is then rapidly made. This should result in instantaneous volatilization of all sample components. The use of an autoinjector in addition to allowing unattended operation also ensures very accurate and reproducible injections.

E Choice of carrier gas

Packed columns are usually operated with nitrogen as the carrier gas although helium is sometimes used. Nitrogen is not a good choice for capillary use as optimum efficiency is attained only at low linear velocities, leading to long analysis times with efficiency falling rapidly as gas velocity is increased. Both helium and hydrogen produce maximum chromatographic efficiency at higher linear velocities (typically between 20 and 30 cm s^{-1} for helium, slightly higher for hydrogen). Moreover, since with these gases column efficiency only decreases slowly as the velocity is increased, flow rates considerably faster than the optimum can be used, thus allowing shorter analysis times without serious loss of separating efficiency.

Carrier gas lines should always be fitted with oxygen filters immediately upstream of the chromatograph, as many liquid phases are rapidly destroyed by oxygen, even when present at low concentrations.

Carrier gas flow rate may be measured by means of a low volume bubble flow meter, however it is simpler and more convenient to measure the linear velocity by injection of methane or a similar unretained solute. Such measurements may be made more accurately and it is not necessary to turn off the detector gases. Moreover inspection of the methane peak affords valuable information about system performance (Jennings, 1981). The linear velocity in cm s^{-1} ($\bar{u}$) is given by the expression $\bar{u} = L/t_m$ where L is the column length in centimetres and t_m is the retention time of the methane peak (in seconds). For those who prefer to think in volume flow rates the corresponding expression for the flow rate (F) in ml min^{-1} is $F = r^2L/t_m$ where r and L are the column radius and length respectively (in centimetres) and t_m is the retention time of the methane peak (in minutes). A suitable carrier gas flow rate is best determined experimentally for a given column and chromatograph, but for a 0.2 mm diameter column using helium as carrier gas a suitable value will probably be in the region 0.5–1.5 ml min^{-1} corresponding to linear velocities of approximately 25–75 cm s^{-1}.

F Detectors

The function of a gas-chromatographic detector is to convert a flow of chemicals into an electrical signal that can be amplified and recorded. The most commonly used is the hydrogen flame ionization detector (FID).

The response of a FID is linear over several orders of magnitude of sample concentration. It will respond to nearly all organic compounds with similar sensitivity (formic acid is the most commonly encountered exception), but is insensitive to impurities such as water and CO_2. The baseline is extremely stable because the response is low in the absence of sample, and the detector is relatively unaffected by small variations in flow rate, pressure or temperature.

Thermal conductivity detectors are occasionally found in microbiology laboratories. While these are less sensitive than FID, they do respond to formic acid, which is useful for analysis of volatile products of metabolism.

G Summary of GLC conditions for analysis of bacterial fatty acid methyl esters

The precise experimental conditions will depend on the equipment and other facilities available, but as a general guide the conditions used in our laboratory are given below.

We use a 12 m × 0.2 mm vitreous silica column, coated with a 0.25 μm film of BP-1 stationary phase (SGE Ltd). This is installed in a Carlo-Erba 4130 chromatograph, which is operated with helium as carrier gas, at a flow rate of 1 ml min^{-1}. The injector and detector oven are maintained at 250 °C.

Both split and splitless injection are used as appropriate. For splitless injection the sample is dissolved in 2–5 ml iso-octane (depending on sample concentration), and 1 μl is injected. The initial oven temperature is 80 °C, and is maintained for 1 minute from injection after which the temperature is programmed to 195 °C and held there for 15 minutes. The split valve is opened (split ratio 1:30) after 30 seconds. For split injection the sample is dissolved in 0.2–0.5 ml hexane and 0.5 μl is injected. The column oven is operated isothermally at 190 °C, and the split valve is permanently open at a split ratio of 1:30. Flame ionization detection is employed in all cases. For packed column operation we have used Pye Unicam 204 and Varian Aerograph 2700 chromatographs, with either nitrogen or helium carrier gas at 30 ml min^{-1}.

Columns used include 3% SE30 on 100/120 mesh supelcoport (in a 2 m × 2 mm silanized glass column), 3% SP2100 DOH (on 100/120 supelcoport packed in 3 m × 4 mm silanized glass column), and 5% DEGS (on Chromosorb W, in a 2 m × 2 mm silanized glass column). The injector and detector temperatures are held at 250 °C and 270 °C respectively, while the column oven is programmed from 160–260 °C at 4 °C min^{-1} and maintained at 270 °C for 10 minutes (for SE30 and SP2100 columns), or is operated isothermally at 180 °C (for DEGS columns). The sample is injected directly on to the column packing in all cases.

III Preparation of fatty acid methyl esters

In intact bacterial cells fatty acids occur chiefly in chemically bound forms as constituents of phospholipids, lipopolysaccharides and other cellular components. It is necessary therefore to use appropriate chemical procedures to liberate them. However, the resulting free fatty acids are too polar and involatile for satisfactory gas-chromatographic analysis, and must first be converted into volatile derivatives. Methyl esters are most usually chosen for this purpose as they are easily prepared, have excellent chromatographic properties and good long-term stability. Moreover release of bound fatty acids and methyl ester formation can be achieved in a single procedure.

Most of the available methods are based on either an alkaline saponification or acid solvolysis. The disadvantages of alkaline saponification are: firstly, the release of amide linked acids may be incomplete (Jantzen *et al.*, 1978); and secondly, unsaturated compounds may be artefactually produced from *O*-substituted 3(OH) acids (Rietschel *et al.*, 1972). Hydroxylated acids however are present in legionellae only as very minor constituents. This procedure can be applied to wet cells, thus eliminating the need for freeze-drying which many will consider an advantage.

The main objection to procedures involving acid methanolysis is that cyclopropane acids may be degraded, with the formation of methoxy products (Lambert & Moss, 1983; Vulliet *et al.*, 1974). Since cyclopropane 17 is a component of *Legionella*, and varies in amount between species, it is important that it is not destroyed in the course of the analysis. In this laboratory we routinely perform transesterification in acidified methanol at 50 °C overnight (30 °C cooler than most published methods). This procedure gives quantitative results that are essentially identical to the alkaline saponification method of Moss and colleagues (1974) and mass spectrometry has not produced any evidence of artefact formation. Hence it would seem that degradation of cyclopropane acids is a serious problem only at high temperatures.

Protocols are given below for both alkaline saponification and acid methanolysis: comparable results should be obtained with either, so choice of method will depend on personal preference and the facilities available.

A Acid methanolysis

1. Cells are carefully scraped from the surface of BCYE agar plates and placed in a tube (suitable 5 ml screw-cap vials are available from Chromacol

Ltd). Typically growth from a single plate after 48–96 hours incubation is adequate. If possible the growth time should be standardized to improve reproducibility.

2. The cells are freeze dried, and 1 ml of toluene is added, followed by 1 ml of methanol, acidified with 50 μl of concentrated sulphuric acid. The acid should be added to the methanol and thoroughly mixed before addition to the cells to prevent charring. If available 2 M HCl in methanol (anhydrous) may be substituted for the sulphuric acid/methanol reagent (Jantzen & Bryn, 1985). The tubes are sealed with PTFE-lined screw caps, and incubated at 50 °C overnight in a heating block or water bath. The caps should be checked for tightness when the tubes are at temperature.
3. The tubes are removed from the heating block, allowed to cool, and 1 ml of a saturated solution of NaCl is added. This is followed by 2 ml of hexane:chloroform (4:1), after which the tubes are vigorously agitated using a vortex mixer (minimum time 30 seconds) to extract the fatty acid methyl esters into the hexane phase. After the organic and aqueous layers have separated the organic (upper) layer is carefully removed with a Pasteur pipette and the extraction is repeated with a second 2 ml portion of hexane:chloroform. If the phases do not separate, a short centrifugation will normally break the emulsion. The two hexane extracts are pooled and concentrated to about 1 ml in a stream of dry nitrogen.
4. Aqueous 0.3 M NaOH (3 ml) is added to the extract, it is mixed and the organic layer removed to a fresh vial. This step, introduced by Miller (1982), removes residual acid and so reduces deterioration of the column.
5. The extract is evaporated to dryness in a stream of nitrogen, and redissolved in 1–5 ml of iso-octane prior to gas chromatography. Alternatively the extracts may be stored dry at −20 °C until required for analysis.

B **Alkaline saponification procedure** (Moss *et al.*, 1974; Lambert & Moss, 1983)

1. Sterile distilled water (0.5 ml) is added to the surface of a BCYE agar plate, the cells are gently scraped off and the suspension transferred to a 125 × 16 mm test tube with a PTFE-lined screw cap (available from BDH Ltd).
2. 2 ml of 5% NaOH in 50% aqueous methanol is added and the tube sealed with a PTFE-lined screw cap. After heating for 30 minutes at 100 °C the tube is removed and allowed to cool to room temperature.
3. The contents of the tube are adjusted to pH 2 by the dropwise addition of approximately 0.5 ml of 6M HCl and, after thorough mixing, checked with pH indicator paper.
4. Boron trichloride/methanol (2 ml of a 10% solution available from Sigma Chemical Co.) is added and the mixture heated for 5 minutes at 85 °C.

5. The tubes are then cooled and extracted twice with 5 ml portions of hexane:chloroform (4:1), the extracts are pooled and concentrated in a stream of dry nitrogen.
6. The extracts are washed with 3 ml of 0.3 M aqueous NaOH as above, dried and dissolved in iso-octane for gas chromatography.

IV Interpretation of results

A Confirmation of identity of fatty acid methyl esters

Standard mixtures of bacterial FA are available from several suppliers. All the major FA occurring in legionellae except i14:0 are present in 'Bacterial acid methyl esters mix CP' (Supelco Ltd). Preliminary tentative identifications may be made by comparison of relative retention times (calculated relative to n16:0).

Unsaturated FA can be distinguished from saturated acids by catalytic hydrogenation. A suitable procedure is as follows:

1 After initial GLC analysis the remaining sample is transferred to a 5 ml screw-cap vial and the solvent evaporated in a stream of nitrogen.
2. Chloroform:methanol (1 ml of a 2:1 mixture) is added, together with a small teflon coated magnetic stirrer, and approximately 5 mg of 5% palladium on charcoal (BDH Ltd). The vial is then sealed with a screw-cap Teflon faced silicone rubber septum.
3. The rubber septum is pierced with two hypodermic syringe needles, the tube is flushed with hydrogen and stirred for 15 minutes at room temperature.
4. Step 3 is repeated twice with additional hydrogen.
5. The catalyst is removed by filtering the chloroform:methanol mixture through a Pasteur pipette plugged with glass wool.
6. The chloroform:methanol is evaporated off, the sample redissolved in an appropriate solvent, and the chromatography is repeated without alteration of the original conditions.

Unsaturated FA will be saturated by the above procedure and on non-polar columns will exhibit increased retention times. Saturated FA and cyclopropane acids will be unaffected.

Ideally, the identities of peaks should be confirmed by gas chromatography/mass spectrometry. Ante-iso branched FA are easily distinguished from iso and unbranched FA by the intensity ratios for the peaks at M-29 and M-31 (M-29 > M-31 for ante-iso; Ryhage & Stenhagen, 1960). However, methyl esters of iso and normal FA produce nearly identical mass spectra. Similarly unsaturated and cyclopropane acids cannot be distinguished, nor can the positions of the double bonds and cyclopropane rings be established.

Preparation of picolinyl esters, followed by mass spectrometry allows structure confirmation of all the above compounds in a single experiment (Harvey, 1982). A procedure suitable for bacterial FA is given in Wait and Hudson (1985). Using these techniques we have demonstrated that the n16:1 acid is always 9-hexadecenoic acid, and that the cyclopropane 17 (cyc 17) is 9,10-methylene hexadecanoic acid (Wait *et al.*, 1986).

B Reproducibility

There are three potential causes of variation in the FA profiles of *Legionella* species. These are: firstly, the intrinsic strain-to-strain variation, the extent of which appears to differ from species to species; secondly, alterations of FA composition due to variations in growth conditions (media, length of culture, temperature, etc.); and, finally, differences arising from the analytical procedure itself (i.e. from the extraction and chromatography).

If all strains are routinely grown on BCYE agar plates, there should be little variation in FA composition that is attributable to the media. However, care should be taken when harvesting cells to ensure that the plate surface is not damaged and that no particles of agar are taken for analysis along with the legionellae. Moss and Dees (1979a) compared the FA compositions of 36 strains of *L.pneumophila* after growth on different media, and found that the profiles were essentially identical on each.

In the cases of at least some species the age of cultures can have a perceptible influence on FA composition. Thus Moss and colleagues (1983) observed that *L.feeleii* grown for 24 hours produced 37% n16:0 and only 7% i16:0, while at 72 hours the proportions were 17% and 21% respectively. This behaviour does not appear to be exhibited by *L.pneumophila*, but it is clear that growth times should be standardized if valid comparisons of data are to be made.

The concentration of cyc 17 seems to be particularly sensitive to culture conditions. This acid is synthesized by an enzyme catalysed methylation across the double bond of 9-hexadecenoic acid. Hence levels of the two compounds tend to be reciprocally related. There is evidence that synthesis of cyclopropane acids is a function of the bacterial growth cycle, being low in log phase but accelerating during post-exponential growth (Halper & Norton, 1975). That this is also the case for legionellae, is supported by the observation that when *L.pneumophila* is grown in chemostat culture (in which nearly all cells are in log phase) production of cyc 17 is very low (Figure 7.2; Dennis & Wait, unpublished data). It therefore follows that for organisms grown on plates the proportion of cyclopropane acids present will increase with the age of the culture, as there will be fewer cells in log-phase. This variability of cyc 17 synthesis may explain some of the anomalies in the published data; for example the differing reports of the quantity of n16:1 in

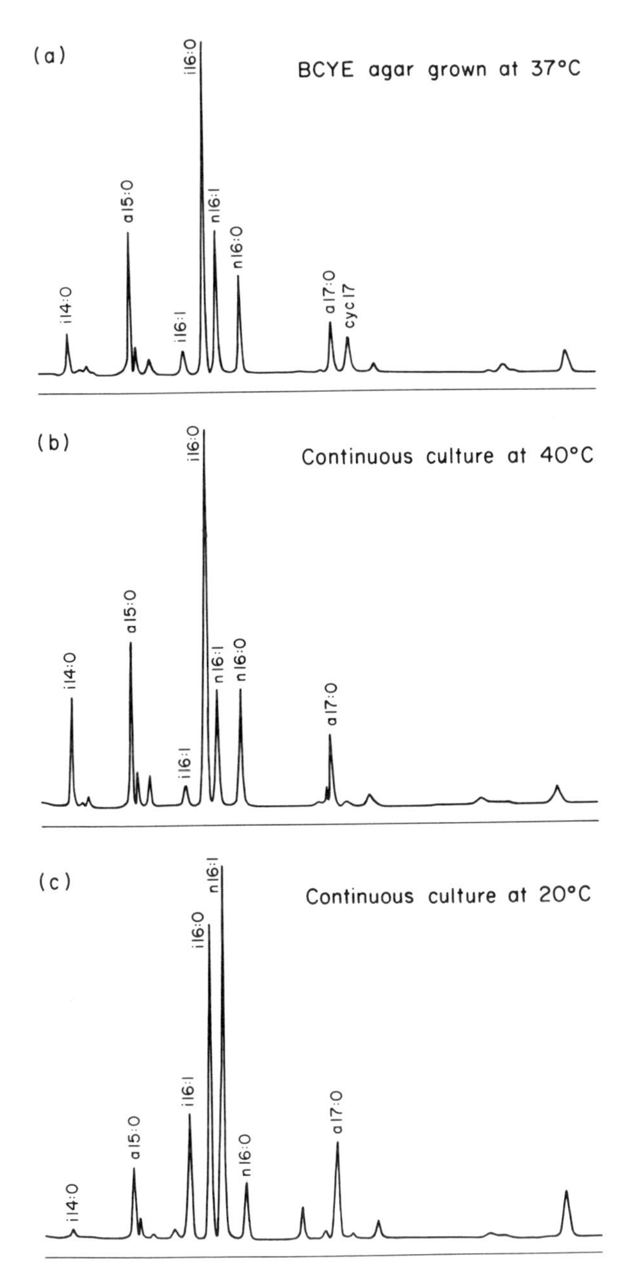

Figure 7.2 Effect of temperature on the fatty-acid profile of *L.pneumophila*.

L.longbeachae and related species (see below). It can therefore sometimes be useful to calculate the total percentage of n16:1 and cyc 17 when making comparisons between these strains.

The temperature at which organisms are grown can also have a dramatic effect on their FA composition. *L.pneumophila* grown in a chemostat at 40 °C exhibits a very similar FA composition to that typically observed for *L.pneumophila* grown on BCYE plates. However, as the temperature is reduced below 30 °C the proportion of unsaturated FA increases until at 20 °C n16:1 is the most abundant FA present and the 'characteristic' profile of *L.pneumophila* has been abolished entirely (Figure 7.2). Hence culture should always be performed at a standard temperature between 37 °C and 40 °C to maximize synthesis of branched chain acids.

The extraction and chromatographic procedures should not themselves contribute to poor reproducibility. This is demonstrated by the data shown in Table 7.1, which show the FA composition of five isolates of '*L.nautarum*' (Dennis *et al.*, unpublished). Although the analyses were performed on different occasions the profiles are very similar. This stability is probably due in part to the absence of cyc 17; other species may show greater variability (e.g. *L.bozemanii*), although whether this is due to greater intrinsic variation in FA composition, or to greater sensitivity to conditions of culture is not clear. However, even when different strains of the same species exhibit differences in their percentage composition, this does not usually have any significant effect on the visual appearance of the chromatogram.

Table 7.1 Reproducibility of the fatty acid profile of '*L.nautarum*'. The data presented were obtained from five different strains determined on different occasions.

	Total cellular fatty acids (%)									
Organism	i14:0	a15:0	i16:1	i16:0	n16:1	n16:0	a17:0	cyc 17	n17:0	n18:0
'*L.nautarum*'*1472*	<1	5	1	11	17	14	21	—	10	8
'*L.nautarum*'*1477*	<1	5	1	11	19	15	23	—	8	8
'*L.nautarum*'*1475*	<1	6	1	13	17	12	24	—	8	6
'*L.nautarum*'*1476*	<1	5	<1	10	15	11	30	—	14	13
'*L.nautarum*'*1471*	<1	5	1	13	17	13	23	—	10	7

C Fatty acid profiles of *Legionella* species

Figure 7.3 shows FA profiles of the currently recognized species of *Legionella*.

L.pneumophila contains i16:0 as its major FA with lower amounts of i14:0, a15:0, a17:0, cyc 17, n16:0 and n16:1. In common with all other *Legionella* species there are only low levels of hydroxy acids (Mayberry, 1981, 1984). It has been suggested that *L.pneumophila* strains may be subdivided into three groups on the basis of differences in FA profiles (Moyer *et al.*, 1984), but this observation has not been confirmed by any other workers. It seems likely that the unidentified peaks observed by Moyer and colleagues were artefacts of derivatization. We have examined isolates representative of nine serogroups of *L.pneumophila* and have not observed any important or consistent differences between them.

The species with a FA profile most similar to *L.pneumophila* is *L.longbeachae* which also has i16:0 as the dominant acid, although usually the levels of n16:1 and a15:0 are higher than in *L.pneumophila*. Moss and colleagues (1981) published data in which n16:1 is the major FA in several strains of *L.longbeachae*, although Mayberry (1984) also reports i16:0 as the dominant acid. It is probably significant that we observe increased cyc 17 as well as decreased n16:1 since, as shown above, levels of the two acids tend to be inversely related and this may account for the observed variability. In any event higher levels of n16:0 and a15:0 usually allow *L.longbeachae* to be distinguished from *L.pneumophila*. The isoprenoid quinone profiles of the two species are unequivocably different, and so should resolve any remaining ambiguity (see below).

L.sainthelensi has a very similar profile to *L.longbeachae*, again having a relatively high proportion of n16:1 which, in some strains, may be more abundant than i16:0 (Campbell *et al.*, 1984). The similarity of the profile to that of *L.longbeachae* is not surprising, as the two organisms are DNA related at the 37% level (Brenner *et al.*, 1985; Campbell *et al.*, 1984). *L.santicrucis*, which is related to both *L.longbeachae* and *L.sainthelensi* is very similar in its FA composition to these species having i16:0 as the major FA. The one strain we have been able to examine also contained high levels of cyc 17 but this may not be a consistent feature of the species.

The FA profile of *L.spiritensis* could also possibly be confused with that of *L.pneumophila*. However, it contains less i14:0 ($<2\%$) than is observed in *L.pneumophila* (5% typically) and larger amounts of i16:1. The proportion of a17:0 is also higher, and cyc 17 is present only at a trace level. The most decisive difference is the presence of a17:1, which occurs in *L.pneumophila* as a very minor component ($<0.5\%$).

Although *L.oakridgensis* contains i16:0 as its major FA, it is unlikely to be confused with *L.pneumophila* or any other species: i14:0 is present in trace amounts only, a15:0 and a17:0 at approximately 2% and 5% respectively, and the concentrations of a15:0 and n15:1 are almost equal. Cyclopropane 17 is much more abundant than a17:0 (10–15% of total acids), and most unusually there are quite large amounts of n18:0 ($>10\%$) (Orrison *et al.*, 1983b). There are differing reports in the literature of the amount of cyc 17 in *L.oakridgensis*. Orrison and colleagues (1983b) found about 9%, but Lambert and Moss (1983) detected almost double this (17%). It is noteworthy that the former group of workers also found higher levels of n16:1 than Lambert and Moss, and that the total of n16:1 and cyc 17 was 27% in both studies. This is in good agreement with the value of 25–30% that we usually find in this laboratory. It would seem therefore that the

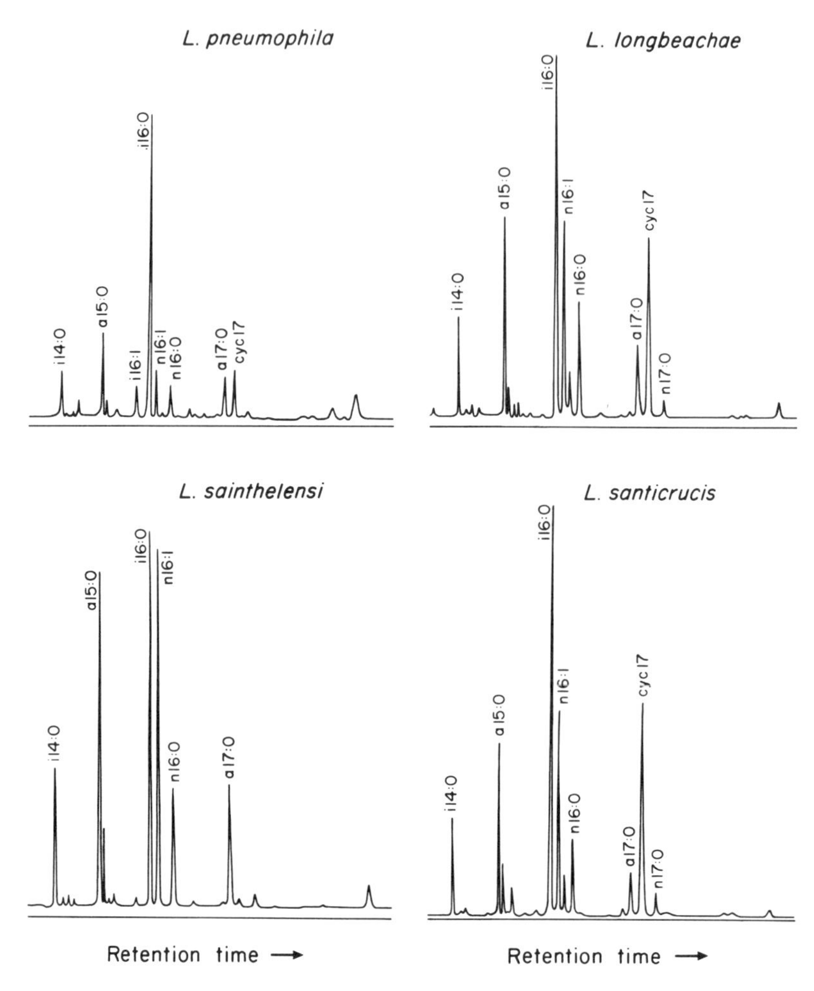

Figure 7.3 The fatty-acid profiles of the type strains of the 28 *Legionella* species.

control of cyc 17 synthesis is especially variable in some species and so caution should be exercised when evaluating the significance of high levels of this acid or its precursor n16:1.

The seven blue/white autofluorescent species; *L.bozemanii*, *L.dumoffii*, *L.gormanii*, *L.anisa*, *L.cherrii*, *L.steigerwaltii* and *L.parisiensis* are ⩾40% DNA related and cannot be distinguished as their profiles are very similar. The profiles are, however, distinct from those of the species described above (Moss and Dees,

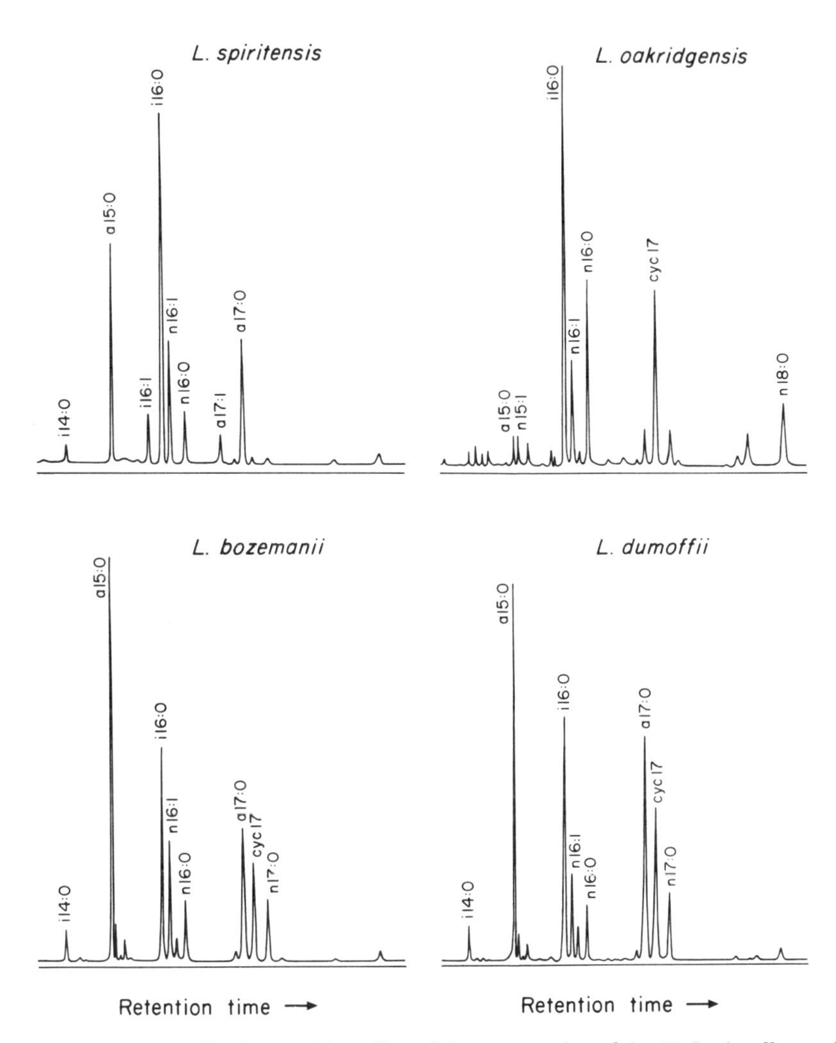

Figure 7.3 contd. The fatty-acid profiles of the type strains of the 28 *Legionella* species.

1979b; Gorman *et al.*, 1985). The profiles all have a15:0 as the major FA (25–30%) with i16:0 as the next most abundant (15–20%), with rather less a17:0 and cyc 17 (typically a17 > cyc 17). Only traces of i16:1 are observed. As would be expected the FA composition of *L. bozemanii* serogroup 2 is not distinguishable from that of serogroup 1, or the other members of the group (Tang *et al.*, 1984).

L. hackeliae and *L. israelensis* have very similar FA compositions to those of the blue/white autofluorescent species, but differ in that i16:1 is present in greater

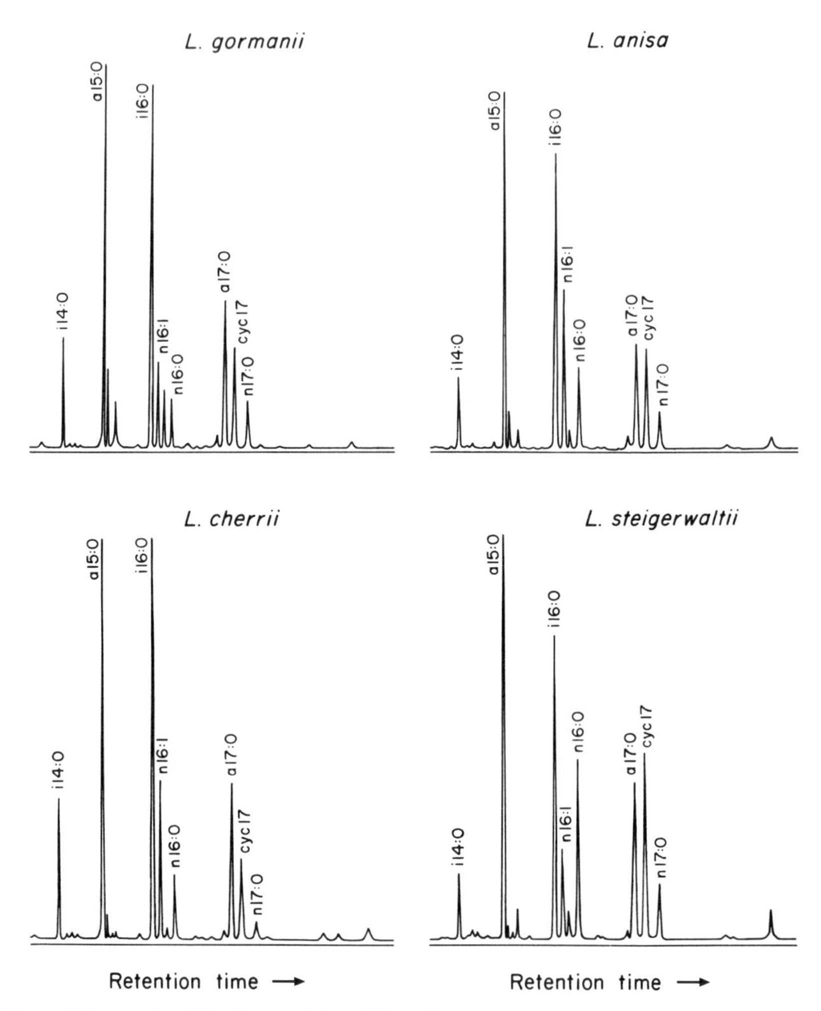

Figure 7.3 contd. The fatty-acid profiles of the type strains of the 28 *Legionella* species.

amount. Some workers may feel that this feature is a rather weak basis for distinction. It is doubtful that the two species can be distinguished apart by their FA content.

L.micdadei also has a15:0 as its major FA (30–40%), but is readily distinguishable from the blue/white autofluorescent species because a17:0 is more abundant than i16:0, and cyc 17 is present only in low amounts. A further characteristic feature is the presence of a17:1 (>5%), an acid which is usually

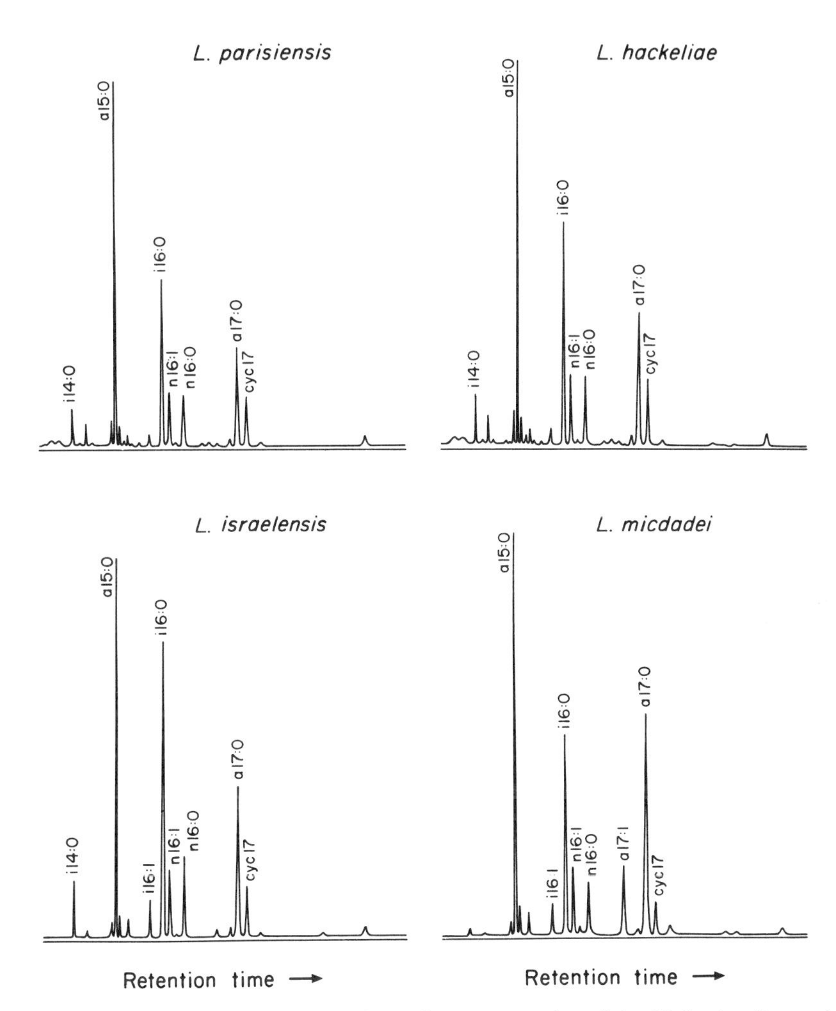

Figure 7.3 contd. The fatty-acid profiles of the type strains of the 28 *Legionella* species.

found (if at all) in trace amount in most other species. *L.maceachernii* has high DNA homology with *L.micdadei* (Brenner *et al.*, 1985) and has a similar FA composition. It too contains a17:1, and in comparable amounts. Thus these two species are probably not distinguishable on the basis of their FA profiles alone. However the only strain of *L.maceachernii* available to us appeared to have slightly more n16:1 than i16:0, whereas the reverse is the case in *L.micdadei*.

Several other species have profiles which are similar to, but distinguishable from *L.micdadei*. *L.jordanis* (Cherry *et al.*, 1982) contains a similar proportion of

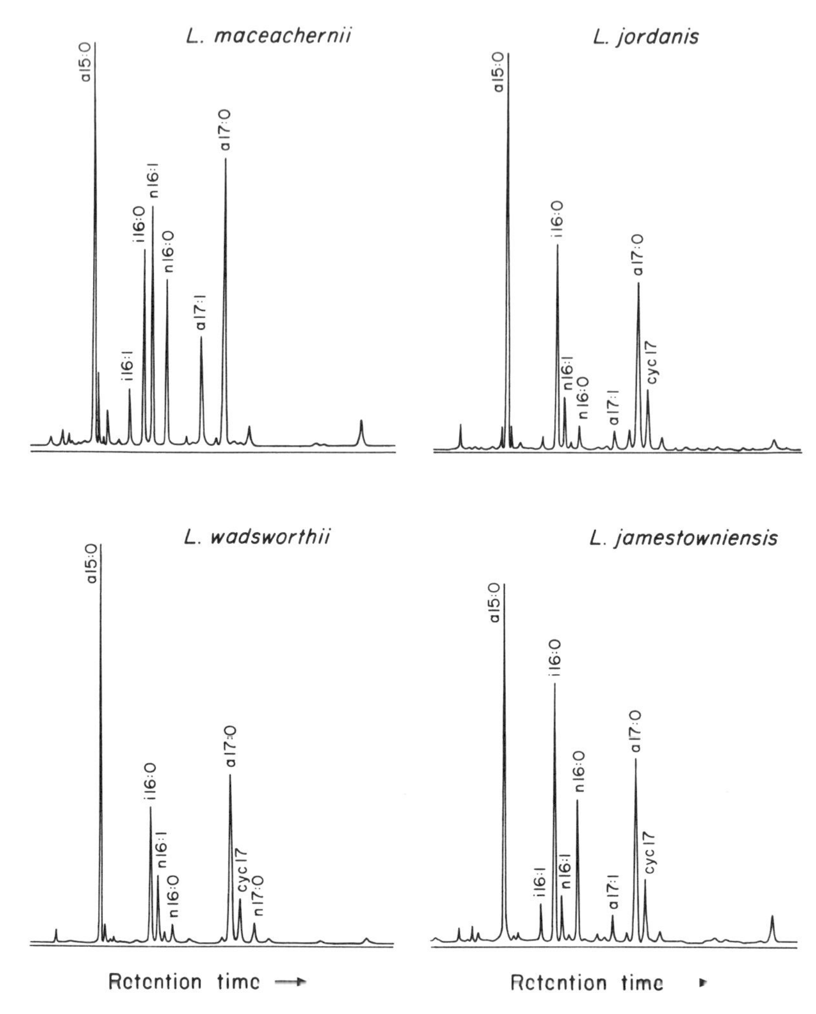

Figure 7.3 contd. The fatty-acid profiles of the type strains of the 28 *Legionella* species.

a15:0 (40%), but the integrated area of a17:0 is only slightly greater than that of i16:0. Also a17:1 is present but in about one-third of the quantity occurring in *L.micdadei*. *L.wadsworthii* also has a similar FA profile to that of *L.micdadei* (Edelstein *et al.*, 1982a) with more a17:0 than i16:0, but can be easily distinguished by the absence of a17:1 (and also of i16:1, although this is only a trace component in many species).

L.jamestowniensis appears to be intermediate in its FA composition between *L.bozemanii* and *L.micdadei*. It can be distinguished from *L.bozemanii* and other

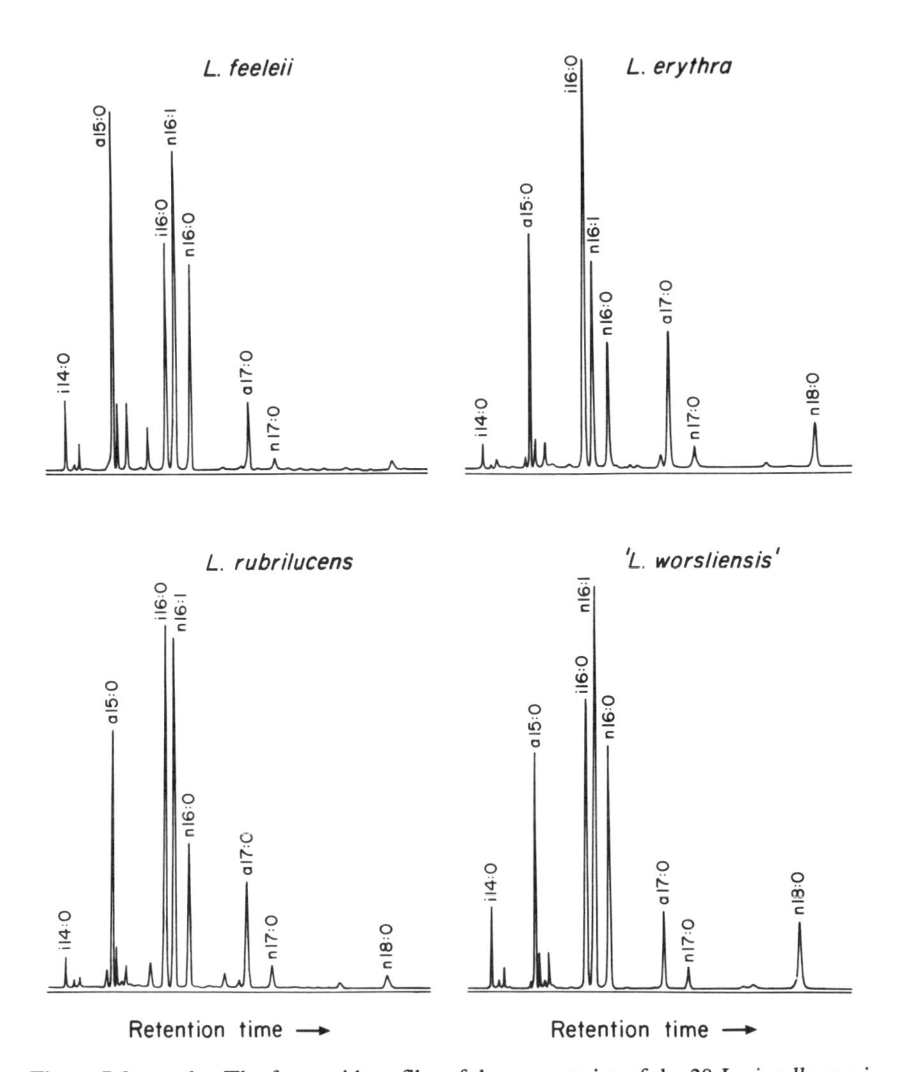

Figure 7.3 contd. The fatty-acid profiles of the type strains of the 28 *Legionella* species.

blue/white autofluorescent species by the presence of greater quantities of both i16:1 and a17:1 (2% of each), and from *L.micdadei* because of its lower proportion of a15:0, together with approximately similar amounts of i16:0 and a17:0.

L.feeleii is another species with a FA profile which is unlikely to be confused with those of other legionellae. The acids a15:0, i16:0, n16:1 and n16:0 are present in approximately equal amounts, although there is usually slightly more a15:0 and n16:1 than i16:0. Levels of a17:0 are low (5%) and there is a complete

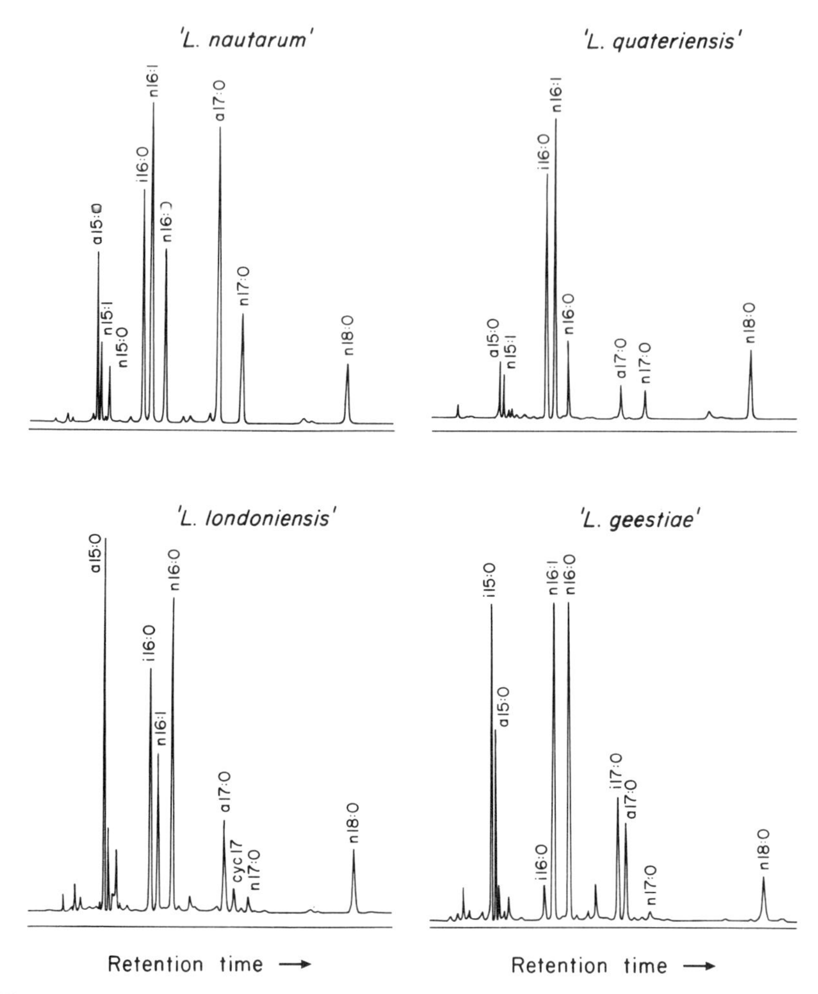

Figure 7.3 contd. The fatty-acid profiles of the type strains of the 28 *Legionella* species.

absence of cyc 17. As mentioned above the proportion of n16:0 is dependent on the age of the cells, falling as the culture gets older (Moss *et al.*, 1983).

The two red autofluorescent species, *L.erythra* and *L.rubrilucens*, are similar in FA composition to *L.feeleii*, having quite high levels of n16:1 and n16:0, although i16:0 is usually the most abundant acid. Cyclopropane 17 is invariably absent in both species, and unsaturated branched acids (e.g. i16:1 and i17:1) are present only in trace amounts. Since the relative proportions of the 16 carbon acids are variable (especially n16:1 and n16:0), it is doubtful if these two closely related species can be distinguished by their FA profiles.

Although *Legionella* species generally contain predominantly branched chain FA, a few recently identified species (Dennis *et al.*, unpublished) contain an unusually high proportion of unbranched FA. These species have some similarity in FA composition to *L.erythra* and *L.rubrilucens* but they do not autofluoresce and hence confusion is unlikely. All lack cyc 17 or contain it only as a trace component. The major FA is usually n16:1 or n16:0 rather than i16:0. In addition n18:0 is present in greater amounts (>5%) than is usual in most other *Legionella* species. The FA profile of '*L.worsliensis*' is similar to both *L.feeleii* and *L.erythra*, although it possesses less a15:0 and more n18:0 than the former, and more i14:0 than the latter. '*L.nautarum*' may be distinguished from other members of this group of new species by the presence of a17:0 as the major FA (24%, Table 7.1). The FA profiles of '*L.quateriensis*' and the unnamed species '1862' (data not shown) are similar to each other in having i16:0 and n16:1 as the major FAs, and fairly low amounts of n16:0 (<10%). Both species are unusual in having approximately equal (but low) amounts of a15:0 and n15:1 (a feature shared with *L.oakridgensis*). The concentrations of a17:0 and n17:0 are also approximately equal, a feature which is not observed in any other species. '*L.londoniensis*' differs from the other new species in having more a15:0, and also traces of cyc 17. Were it not for this latter point it would be hard to distinguish from *L.feeleii*.

Finally another new species, '*L.geestiae*' is distinct from all other *Legionella* species so far described in having significant amounts of odd numbered iso branched acids; in fact both i15:0 and i17:0 are more abundant than their ante-iso branched isomers. Like the previous species, n16:0 and n16:1 are the major FAs, and again there is no cyc 17.

Table 7.2 is an attempt to group the currently recognized *Legionella* species according to the similarity of their FA profiles. Within some of these groups (A, D, E) sufficiently consistent differences exist to allow further subdivisions. While it is true that genetically related species tend to have similar profiles (e.g. the 'Fluoribacter' group) the converse is definitely not the case, and some unrelated species are effectively indistinguishable in FA composition.

V Isoprenoid quinone profiling

As described above all members of the genus have qualitatively similar FA profiles which are, in some cases, quantitatively distinct. Some species, however, cannot always be differentiated on this basis. These include *L.pneumophila* and *L.longbeachae*, which differ only in their relative content of i16:0 and n16:1, and also other closely related species such as the blue/white autofluorescent species. Fortunately the Legionellaceae possess an unusual pattern of respiratory quinones, which allow further discrimination to be made between some species.

Isoprenoid quinones are constituents of bacterial plasma membranes, where they play an important role in respiratory electron transport. The most common classes of bacterial respiratory quinones are the menaquinones (based on a 2-methyl-1,4-napthoquinone nucleus), and the ubiquinones (based on a substituted benzoquinone nucleus). The sidechain may contain 1 to 15 isoprenyl units (Figure 7.4). A comprehensive review of bacterial quinones and their taxonomic value may be found in Collins and Jones (1981a).

Table 7.2 Similarity groups of fatty acid profiles of *Legionella*.

A. 'Pneumophila Group'
(i16:0 major acid)

L.pneumophila
L.longbeachae
L.sainthelensi
L.spiritensis (a17:1 present)
L.santicrucis

B. 'Oakridgensis Group'
(i16:0 major acid; high cyc 17, low a15:0)

L.oakridgensis

C. 'Fluoribacter Group'
(a15:0 major acid; i16:0 > a17:0)

L.bozemanii	*L.anisa*	*L.parisiensis*
L.gormanii	*L.steigerwaltii*	
L.dumoffii	*L.cherrii*	

(*L.israelensis* *L.hackeliae*) i16:1 present

D. 'Micdadei Group'
(a15:0 major acid; a17.0 > i16:0)

L.micdadei } Significant
L.maceachernii } a17:1
L.jordanis
L.wadsworthii
L.jamestowniensis

E. 'Feeleii Group'
(no cyc 17; major n16:0 or n16:1)

L.feeleii (high a15:0)
L.erythra
L.rubrilucens
'*L.worsliensis*'
'*L.nautarum*' (high a17:0)
'*L.quateriensis*' } a15:0 = n15:1
'Species 1862' } a17:0 = n17:0
'*L.londoniensis*' (cyc 17 present)

F. 'Geestiae Group'
(significant i15:0 and i17:0; no cyc 17; high n16:0 and n16:1)

'*L.geestiae*'

Finnerty and colleagues (1979) reported that quinones of the ubiquinone series alone are present in *L.pneumophila* and they demonstrated the presence of a novel component with more than ten isoprene units which, however, they did not characterize. Using thin layer chromatography (TLC) and mass spectrometry Karr and colleagues (1982) were able to identify Q11, Q12, and Q13. These observations were extended by Collins and Gilbart (1983), who demonstrated the presence of Q14 and Q15, and obtained quantitative profiles of the species by high performance liquid chromatography (HPLC).

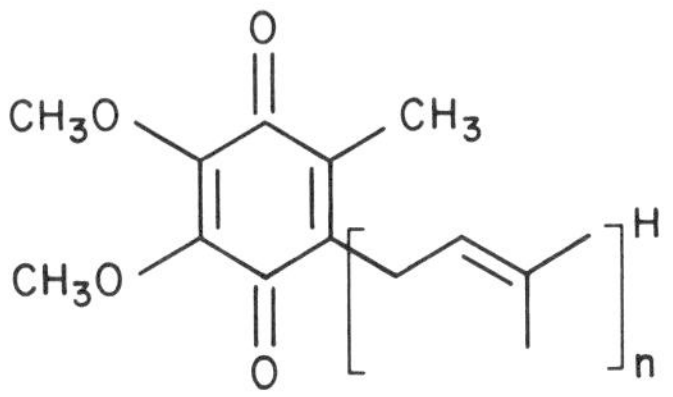

Figure 7.4 Structure of ubiquinones (Q–n). In *Legionella* spp. Q = 6–15.

A Extraction and purification of ubiquinones

Quinones are more delicate molecules than fatty acids and are susceptible to photo-oxidation. It is therefore desirable that they be protected from strong light during extraction and analysis (e.g. by surrounding flasks with aluminium foil), and that analyses are performed as soon as possible after extraction. Water should also be rigorously excluded from reagents used in the extraction procedures as its presence may result in poor recovery of ubiquinones. A number of gentle methods for the extraction of bacterial quinones have been reported (Krivankova & Dadak, 1980; Collins *et al.*, 1977; Collins, 1985; Tamaoka, 1986). Most of these methods are based on extraction with a lipid solvent followed by some form of chromatographic purification, usually on silica. A method for the combined extraction of quinones and polar lipids has also been described (Minnikin *et al.*, 1984).

The following technique, modified from that of Collins and colleagues (1977), is used routinely in this laboratory and has proved very satisfactory for the extraction of quinones from legionellae:

1. Freeze dried cells (20–50 mg) are suspended in 20 ml of chloroform:methanol (2:1) in an aluminium foil covered 50 ml conical flask. A PTFE coated magnetic stirrer bar is added, and the suspension stirred at room temperature for two hours.

2. The bacterial suspension is then filtered through Whatman No. 1 filter paper (12.5 cm). The material retained on the filter paper is washed with a further 10 ml of chloroform:methanol and the combined filtrates are dried by rotary evaporation. The temperature of the water bath should be below 40 °C.
3. The resulting residue is dissolved in a minimal volume of chloroform and applied to a plastic backed silica gel $60F_{254}$ TLC plate (Merck, available from BDH Ltd in the UK). For economy each 20 × 20 cm plate can be cut into four 10 × 10 cm squares. A drawn out Pasteur pipette is used to apply the sample as a narrow band 3–5 cm long approximately 1 cm from one end. If desired a Q10 standard (Sigma) can be spotted on either side of the sample band to aid location of the ubiquinones.
4. The plate is then developed with a petroleum ether (60–80 °C boiling range):diethyl ether mixture (85:15), in a glass TLC tank which has been pre-equilibrated with the solvent, until the solvent front has migrated two-thirds of the way up. The ubiquinone band is then located, and lightly outlined in pencil, by brief examination of the TLC plate under ultraviolet light (254 nm).
5. The area of silica containing the quinones is scraped off the TLC plate, with a scalpel and placed in a sintered glass funnel. If a Q10 standard is being used care must be taken not to collect this with the sample. Chloroform (5 ml) is then used to elute the quinones into an amber coloured glass vial, and the chloroform is evaporated in a stream of nitrogen.

As an alternative to TLC purification, silica 'Sep-paks' (Waters Associates) can be used. These are disposable plastic cartridges, packed with silica, which can be attached to the Luer fitting of a syringe, thus allowing sample to be loaded easily and the eluting solvent to be pumped through. 'Sep-pak' cartridges have been successfully used for purification of both menaquinones (Shearer, 1986) and ubiquinones (Gilbart, 1985). The cartridge is pre-wetted by passing 20 ml of hexane through it. The dried extract from step 2 above is then dissolved in a small volume of hexane and applied to the cartridge by means of a glass syringe: 10 ml of hexane is then pumped through and discarded. The ubiquinone fraction is recovered by elution with hexane:diethyl ether (5:15): the first 4 ml are collected and concentrated under a stream of dry nitrogen.

B Chromatographic analysis of ubiquinones

Reverse-phase high performance liquid chromatography (RPHPLC) is the method of choice for quinone analysis, and essential if quantitative results are to be obtained. However, the required equipment is expensive and often not available in microbiology laboratories. Fortunately reverse-phase thin layer chromatography (RPTLC) is a simple and cost effective alternative, which while not providing quantitative data (unless a scanning densitometer is used), is usually quite adequate for the routine identification of strains.

i High performance liquid chromatography of ubiquinones

In high performance liquid chromatography (HPLC), as in GLC, separation is effected by partitioning sample components between a mobile phase and a stationary phase. In the case of HPLC the mobile phase is a liquid. As separation proceeds at, or close to, ambient temperature polar and thermally unstable compounds may be analysed much more easily than by GLC. Moreover the chromatographer has considerable control over the process of separation as, in addition to selecting a suitable stationary phase, it is also possible to change the mobile phase and even to alter its composition and polarity during the course of the analysis (gradient programming).

Solvent gradient programming is not in fact necessary for the separation of *Legionella* quinones as excellent resolution is obtained with isocratic mixtures of methanol and 1-chlorobutane, using C18 reversed-phase columns. Readers desiring detailed accounts of the equipment required for HPLC should consult one of the excellent texts available on the principles and practice of the technique such as Snyder and Kirkland (1979), Knox (1981), or Lim (1986).

In quinone analysis RPHPLC is used almost exclusively (Moss & Guerrant, 1983; Collins & Gilbart, 1983). In such systems the elution solvent is more polar than the stationary phase. The support material consists of a high quality silica of carefully controlled particle size and pore distribution. The column is rendered non-polar by chemically bonding alkyl silanes to surface SiOH groups, octadecyl silanes being most commonly used (ODS or C18 columns).

The affinity of quinone molecules for such columns is largely determined by the nature of the isoprene chain at carbon 6, which is non-polar so will be strongly absorbed by the column, longer chains tending to be more tightly associated. Since the quinones of *Legionella* species differ only in the length of the isoprene chain, they will be eluted in order of chain length, short chain quinones emerging first, and Q15 last.

Several commercially available C18 columns have been successfully used for ubiquinone analysis. They include Spherisorb ODS (250 × 4.6 mm) (Laboratory Data Control) using methanol:1-chlorobutane (90:10) at 1.5 ml min^{-1} as mobile phase, and μBondpak C18 (300 × 2.0 mm) (Waters Associates) using methanol:1-chlorobutane (80:20) at 0.5 ml min^{-1} (Gilbart, 1985). In addition to these conventional stainless steel columns we have also used a Waters RCM 100 radial compression module with an 8 mm × 10 cm μBondpak C18 radial pak cartridge, with methanol:1-chlorobutane (70:30) at 1.5 ml min^{-1}. In all cases 1-chlorobutane was used as the injection solvent and eluted quinones were detected by ultraviolet absorption at 270 nm.

The system in use in our laboratory consists a pair of M6000A pumps, a U6K manual injection valve, a WISP model 710B automated injection system, with a model 481 variable wavelength UV detector and 730 data module and 720 system controller (all from Waters Associates). However, a minimum system for ubiquinone analysis would consist of a single pump, manual injection valve, column, UV detector and recorder. If a 270 nm UV detector is not available then 254 nm can be used, but with some loss of sensitivity.

ii Confirmation of ubiquinone identity

Standard ubiquinones (Q6–Q10) are commercially available (Sigma). At the time of writing there is no commercial source of Q11–Q15. Hence *Legionella* species of known quinone composition should be used as standards for these compounds.

Ubiquinone mixtures may be identified by mass spectrometry using a direct insertion probe. The base peak is observed at m/z 235, with a further intense peak at m/z 197. Strong molecular ions will be observed at even mass numbers corresponding to the molecular weight of each ubiquinone species present, e.g. m/z 930, 998 and 1066 from Q11, Q12 and Q13 respectively. Mass spectra of individual ubiquinones may be obtained by collecting peaks as they emerge from the UV detector, followed by direct-probe mass spectrometry. UV spectra may also be recorded on the collected peaks: ubiquinones exhibit simple spectra with absorption maxima at 270–275 nm (Collins, 1985).

Finally, if authentic standards of two or more ubiquinones are available then a plot of the logarithm of the adjusted retention time (i.e. elution time of ubiquinone minus elution time of unretained solvent) against number of isoprene units will produce a straight line from which the retention time of other members of the series can be extrapolated (Gilbart & Collins, 1985; Tamaoka, 1986).

iii Reverse-phase thin layer chromatography of ubiquinones

RPTLC may be used in laboratories where HPLC is not available and is also useful for initial screening prior to analysis by HPLC. Commercially available plastic backed plates are used, coated with silica to which octadecyl silane groups have been chemically bonded. Although the plates may seem expensive, up to ten strains may be run on a single 10 × 10 cm plate. Moreover, apart from the plates the only other requirements are a conventional glass TLC tank and a small volume of eluting solvent. Thus compared with HPLC the technique is rapid and cost effective. The following method is based on that of Collins and Jones (1981b):

1. The ubiquinone samples are dissolved in a small volume of acetone and, using a micropipette or microsyringe, carefully spotted on to 10 × 10 cm Merck HPTLC RP $18F_{254}$ plates (available in the UK from BDH Ltd) approximately 1 cm from the lower edge at intervals of 1 cm. In addition to any available ubiquinone standards, a sample obtained from a *Legionella*

species of known quinone composition should be applied to each plate.

2. The plates are developed in acetone:acetonitrile (80:20) until the solvent front is about three-quarters of the plate height, fifteen minutes usually being sufficient.

The order of elution is the same as that seen in reverse-phase HPLC, mobility being inversely proportional to chain length, so the longest quinones remain closest to the origin. The separated ubiquinones may be visualized under ultraviolet light (254 nm), showing up as dark spots against the fluorescent plate. Alternatively the plates may be sprayed with a 5% solution of phosphomolybdic acid in ethanol, followed by heating at 120 °C for a few minutes. The quinones are then revealed as dark blue spots on a pale yellow background. This latter procedure however is destructive, precluding elution and further characterization of the ubiquinones.

C Ubiquinone composition of *Legionella* species

Data on the quinone composition of *Legionella* species examined in this laboratory are given in Table 7.3. Not every species has yet been examined, but those that have all contain quinones with more than ten isoprene units. Since these compounds are highly unusual, their detection in putative legionellae provides strong confirmatory evidence of the identification.

Where different serogroups of the same species have been examined, no significant differences were observed. Thus serogroups 1 to 6 of *L.pneumophila* exhibit profiles which are effectively identical as do serogroups 1 and 2 of *L.longbeachae* (Collins & Gilbart, 1983).

In some cases where two species have similar fatty acid profiles quinone data may help to discriminate between the possibilities. Thus *L.pneumophila* can sometimes be confused with *L.longbeachae* as both have high levels of i16:0 but their quinone profiles are different; *L.pneumophila* having lower levels of Q9, Q10 and Q11, but higher Q12 and Q13. *L.sainthelensi*, however, has a very similar profile to *L.longbeachae*, to which it is closely related, and it is doubtful if the two species can be separated using this technique.

The blue–white autofluorescent species, *L.bozemanii*, *L.anisa*, *L.gormanii* and *L.dumoffii*, are very similar to each other in quinone composition. The profiles are dominated by Q10, Q11 and Q12, with Q12 predominating. *L.dumoffii* and *L.gormanii* appear to have more Q12 (>40%) and less Q10 than the other members of the group, but it is not clear if this difference is sufficiently consistent to be useful in identification.

The quinone profile of *L.micdadei*, containing more than 60% Q13 and about 30% Q12, is more distinct from those of the above species than is its FA acid composition. *L.jordanis*, which is very similar to *L.micdadei* in FA composition, is also virtually indistinguishable in quinone composition. However, *L.wadsworthii*, which is similar to *L.micdadei* and *L.jordanis* in its FA composition, has a very different quinone profile consisting of about 75% Q10, with no detectable quinones beyond Q11.

Table 7.3 Ubiquinone content (%) of some *Legionella* species.*

	Q7	Q8	Q9	Q10	Q11	Q12	Q13	Q14	Q15
L.pneumophila	—	t	2	1	8	61	27	1	—
L.longbeachae	3	3	10	25	25	30	4	—	—
L.sainthelensi	—	t	10	21	29	40	t	—	—
L.oakridgensis	—	—	17	52	20	11	—	—	—
L.dumoffii	—	1	6	12	29	47	6	t	—
L.gormanii	2	3	8	15	19	44	9	t	—
L.bozemanii	1	2	11	25	24	32	5	—	—
L.anisa	—	t	8	26	27	36	3	—	—
L.micdadei	—	—	t	t	1	28	67	4	t
L.jordanis	—	—	t	2	2	32	60	3	t
L.wadsworthii	—	—	14	75	11	—	—	—	—
L.rubrilucens	—	—	—	—	12	77	12	—	—
L.feeleii	—	—	—	—	—	4	58	38	—
'*L.londoniensis*'	—	—	—	—	—	9	82	9	—
'*L.geestiae*'	—	—	—	—	—	2	45	50	3
'*L.quateriensis*'	—	—	—	—	1	27	69	3	—
'*L.nautarum*'	—	—	t	17	68	16	2	t	—

t = trace (<1%)
— = not detected
*Data reproduced from Gilbart (1985) with permission.

The profile of *L.feeleii* is dominated by Q13 (about 60%) but differs from *L.micdadei* in having almost 40% of Q14. This profile is most similar to that of '*L.geestiae*', which also has, to some extent, a similar FA composition.

In general the proposed new species '*L.londoniensis*', '*L.quateriensis*', '*L.nautarum*' and '*L.geestiae*' differ more from each other in their quinone content than they do in their FA profiles. With the exception of '*L.nautarum*' they have rather few quinones, one or two usually predominating, and tend to be dominated by Q12 and higher prenologues. The quinone content of '*L.nautarum*' is strikingly similar to that of '*L.micdadei*', although they are readily differentiated by their FA composition and other biochemical properties.

Although quinone compositions are a valuable adjunct to fatty acid profiles, both for defining members of the genus and as an aid to speciation, the data should be used with caution for this latter purpose, as the reproducibility of the quantitative profiles has not been studied as fully or as systematically as the corresponding fatty acid data. There are, moreover, discrepancies in some of the analyses published by different laboratories. Thus Campbell and colleagues

(1984) reported equal amounts of Q9, Q10, Q11 and Q12 in *L.sainthelensi*, while Gilbart and Collins (1985) observed appreciably less Q9 and Q10. Such minor discrepancies are not significant in practice, but do serve as a warning that relatively small quantitative differences are probably not in themselves a safe basis for designating two strains as different, particularly if comparisons are being made with literature data.

VI Conclusions

None of the methods described above should be seen as alternatives to serological or biochemical identification techniques, and are most sensibly used in conjunction with them. Thus the use of fatty acid and ubiquinone profiles allows all members of the Legionellaceae to be assigned to the genus, even in the case of hitherto undescribed species or serogroups. Furthermore, because the fatty acid profiles can be used to assign strains to 'similarity groups', each containing only a few species, the consequent narrowing of possible identities allows serological and other techniques to be deployed more effectively and selectively, and provides an additional means of confirming such results.

Acknowledgements

I am very grateful to P. J. Dennis and M. J. Hudson for collaboration and much helpful discussion, to G. Vesey for growth of strains, to J. Gilbart for permission to reproduce quinone data, to Mr C. Engel for bibliographical searches and to Miss Theresa Gallagher for typing the manuscript.

A Laboratory Manual for *Legionella*
Edited by T. G. Harrison and A. G. Taylor

CHAPTER 8

Demonstration of Legionellae in Clinical Specimens

T. G. Harrison and A. G. Taylor

I Introduction

The presence of legionellae in clinical specimens can be demonstrated by a variety of methods. The detection of viable organisms by culture is clearly the definitive technique and this has been discussed in detail in Chapters 3 and 4. Other techniques rely on the indirect demonstration of organisms, using bacteriological stains, or on serological methods. These techniques, which should be used in conjunction with culture and isolation methods, can often enable a diagnosis to be established very rapidly. The selective procedures and media used in attempts to isolate legionellae may inhibit the growth of fastidious *Legionella* species or even of *L.pneumophila*, when present in only small numbers, but organisms may nevertheless be visualized by microscopy using appropriate staining procedures. Similarly a diagnosis may be established by microscopy when organisms are no longer viable, after antibiotic therapy, or retrospectively in fixed tissues.

II The demonstration of legionellae using bacteriological stains

L.pneumophila was first seen by McDade and colleagues (1977) in impression smears of infected guinea-pig tissue stained by the Gimenez method (Gimenez, 1964). This method, which was developed to detect rickettsiae, is simple to perform and has been widely used. Legionellae are seen as bright red pleomorphic rods against a blue background. Silver impregnation stains, such as the Dieterle stain (Dieterle, 1927) or a modified Warthin–Starry stain have also been used to reveal legionellae in clinical specimens (Chandler *et al.*, 1977; Dumoff, 1979), and are the most appropriate for examining paraffin-embedded tissue.

Legionellae in clinical material stain poorly by Gram's method and this technique is generally considered to be of little value in the diagnosis of Legionnaires' disease. There are, however, several reports in the literature where legionellae were seen in respiratory tract specimens from immunocompromised patients using this stain (Arnow & Gardner, 1983; Liu & Wright, 1984; Baptiste-Desruisseaux *et al.*, 1985). Arnow and Gardner suggest that 'in an immunocompromised patient when Gram-negative bacteria are seen in deep respiratory secretions and no growth is apparent on routine aerobic culture media after 24 hours, a diagnosis of Legionnaires' disease should be considered.'

The obvious limitation inherent in the use of bacteriological stains is that they may reveal any bacterial species and so a specific diagnosis cannot be established by their use alone. However, in contrast with the serological methods described below, any *Legionella* species may be revealed and not only those for which specific reagents are available. Therefore to maximize the possibility of detecting legionellae bacteriological staining should be used in addition to a serological method.

III The demonstration of legionellae by indirect immunofluorescence

A Introduction

The demonstration of organisms in clinical specimens by immunofluorescence can be an extremely rapid method of diagnosing LD. The method described below for an indirect immunofluorescent test (IFT) has been successfully used in this laboratory for about ten years. The specimen slide is first incubated in the presence of hyperimmune rabbit antisera. If immunoglobulin is bound this is then revealed in a second step using FITC conjugated anti-rabbit immunoglobulin. An alternative technique is a direct immunofluorescent test (DFT) where the FITC has been directly coupled to the rabbit immunoglobulin, permitting the whole test procedure to be completed in one step (Cherry *et al.*, 1978).

The advantage of the DFT compared to the IFT is that it is more rapidly performed. However, the DFT has several disadvantages. The most significant of these is that individual DFT conjugates are required for each serogroup. A second is that because of non-specific background fluorescence and dilution effects only a small number of DFT conjugates can be pooled together in a single reagent: a constraint which does not apply to the same degree in the case of unconjugated antisera (Brown *et al.*, 1984). In addition the IFT is more economical and has been reported to be at least as sensitive as the DFT (Brown *et al.*, 1984).

The majority of *Legionella* infections are caused by *L.pneumophila* serogroup 1. Monovalent direct conjugates prepared against this serogroup can often facilitate the rapid diagnosis of LD in acutely ill patients. Such conjugates can be prepared as described by Hébert and colleagues (1972) and will soon be available from this laboratory.

The detailed methodology for the production of hyperimmune rabbit antisera and the test procedures for the IFT are given in Chapter 6. The only additional point to be noted here is that whereas in the identification of *Legionella* strains a pure culture of the organism is examined, clinical material will often comprise a mixture of unidentified bacterial species. Thus the appropriate test controls, which are described below, must be used if any confidence is to be placed in the results obtained.

B Collection of specimens

Wherever possible attempts should be made to culture legionellae from appropriate patient specimens. Consequently clinical material will often be received unfixed. Respiratory tract specimens such as sputa, transtracheal aspirates, bronchial washings and pleural fluids are all suitable for examination (Table 3.1, page 16). These should be refrigerated upon arrival in the laboratory and examined as soon as possible to avoid growth of contaminating organisms. Lung tissue not required for culture can be fixed satisfactorily by immersion in 10% formalin overnight or stored at $-20\,^{\circ}C$ (or below) and tested when convenient. Fresh or thawed lung tissue should be examined as soon as possible.

Legionellae have also been demonstrated by immunofluorescence in a variety of extra-pulmonary sites. These include: liver, spleen, kidney, brain, gut, wounds and abscesses (Dorman *et al.*, 1980; Cutz *et al.*, 1982; Dournon *et al.*, 1982,

Brabender *et al.*, 1983; Ampel *et al.*, 1985). Where such specimens are available they should also be examined as described below.

C Preparation of specimens

i Respiratory tract specimens

Thin smears should be made to minimize background fluorescence which can obscure any organisms present. In the case of sputa it is important to prepare smears from the actual sputum and to avoid dilution with saliva. Bronchoalveolar lavages and pleural fluids should be centrifuged (>200*g* for 30 minutes) and a smear prepared from the pelleted material. After allowing the smears to air-dry they should be gently heat-fixed.

ii Lung specimens

Both formalin-fixed or unfixed tissue can be treated similarly. The area of greatest consolidation is chosen and, using a sterile scalpel blade, the tissue is cut to give a clean fresh surface. A slurry of cellular tissue is then obtained by scraping the scalpel across the cut surface. Smears are prepared on microscope slides and air-dried and gently heat-fixed. If fresh tissue is being examined smears can be immersed in 10% formalin for 10 minutes before being rinsed in distilled water and air-dried.

Paraffin-embedded tissue sections can be examined for legionellae but all traces of paraffin must be thoroughly removed. The use of material prepared in this way has the advantage of retaining the histological features which may aid the interpretation of results particularly if intracellular organisms are revealed.

D Test procedures

The test procedures used are essentially the same as those described in Chapter 6. The hyperimmune rabbit antisera are prepared in the same manner and are used at the same working dilutions. The volume of antiserum used will depend on the area of smear to be covered but should be adequate to ensure that the smears do not dry during incubation. Incubation and washing procedures are as previously described. Despite these similarities there are a number of important differences and these will be detailed below.

i Range of antisera

With the exception of a small number of well-documented instances (Stout *et al.*, 1982; Rudin *et al.*, 1984; Herwaldt *et al.*, 1984; Joly *et al.*, 1986a) all reported outbreaks of *Legionella* infection have been caused by *L.pneumophila*, and almost all of these by serogroup 1 strains. In addition the majority of sporadic cases are also caused by this serogroup. Thus if specimen material is limited in volume, examination using a monovalent *L.pneumophila* serogroup 1 antiserum is appropriate. Where possible antisera, either monovalent or polyvalent, against the other *L.pneumophila* serogroups should also be used. There is little value in routinely testing specimens for the presence of other species. The increased demand on resources incurred by testing specimens for all *Legionella* species known to cause illness is not justified outside a reference laboratory.

The species presently known to be capable of causing infection in man are shown below:

L.pneumophila – The majority of sporadic cases and outbreaks.
L.micdadei – Sporadic cases and outbreaks in Pittsburgh (Rudin *et al.*, 1984).
L.bozemanii
L.dumoffii – Sporadic cases and a single documented outbreak in Quebec (Joly *et al.*, 1986a).
L.feeleii – Sporadic cases and an outbreak of Pontiac fever (Herwaldt *et al.*, 1984).
L.maceachernii
L.hackeliae
L.jordanis
L.longbeachae
L.sainthelensi
L.wadsworthii
L.gormanii
L.birminghamensis
L.oakridgensis – Implicated by serological evidence only.

With the exception of *L.pneumophila* infections, illnesses caused by these species are almost always confined to immunocompromised patients.

ii Specimen processing

As the presence of even a few fluorescent organisms in a smear examined by IFT may be taken as evidence of infection, it is extremely important to avoid cross-contamination of negative specimens with positive specimens. All glassware and plastics should be free from bacterial contamination and all washing solutions should be membrane filter sterilized (0.2 μm) before use (legionellae are commonly found in laboratory water supplies). Each smear must be processed in an individual container, and separate smears should be prepared for testing with each antiserum. Smears may be made on any suitable type of microscope slide which should be scrupulously clean. PTFE-coated slides with a well size of about

1 cm are very satisfactory. They provide a sufficiently large area to reveal organisms present only in low numbers, but are small enough to be flooded with 25 – 50 μl of antiserum or conjugate. Although some of the specimen is lost during processing, sufficient bacteria and cellular debris usually remain on the slide for microscopy.

iii Control smears

False-positive results can be obtained in the IFT in several ways. Firstly, bacteria in the test specimen may bind immunoglobulin by non-specific mechanisms, e.g. through the presence of Fc-receptors. If this occurs the FITC antiglobulin conjugate will then be bound by the attached immunoglobulin and a false-positive result will be obtained. This problem can be revealed by using a control smear in which 'normal' rabbit serum is substituted for the antiserum. A positive result with the normal serum indicates that the result with the immune serum is not significant.

A control slide must also be processed using PBS in place of antiserum to act as a conjugate control. Some commercial anti-rabbit conjugate will react with human antibodies. Using such conjugates any bacteria in the specimen which are coated with human antibody will fluoresce and thus give rise to false-positive results.

Wherever possible a clinical specimen, previously found to be positive, should be tested as a positive control to check that the test reagents are reacting correctly. If appropriate clinical material is not available a formalin-killed suspension of a laboratory-grown legionella should be used in its place. The most common source of false-positive results is by cross-contamination of test smears with material from the positive control. This will not occur if each smear is processed in a separate container.

iv Examination of slides

It is essential that slides are examined using a high magnification objective lens. A ×40 lens can be used to screen the specimen but a ×63 or ×100 lens should be used to confirm the identity of any fluorescing bacteria. The typical appearance of legionellae in clinical material is shown in Plate III.k,l. The organisms appear as short rods or cocco-bacillary forms, in contrast to the longer and often filamentous forms seen when organisms are grown on laboratory media. Filamentous forms have been seen and isolated from clinical material in the case of patients who have been treated with antibiotics (E. Dournon, personal communication), but this is very rare. In respiratory tract specimens, particularly sputa, the bacteria may be very sparsely distributed. In lung tissue cellular debris is often visible and clusters of fluorescent intracellular bacteria may be seen which is very characteristic of *Legionella* infections (Plate III.k).

E Interpretation

i Diagnostic criteria

The criteria given here have been established by workers at the CDC Atlanta for the interpretation and reporting of results obtained using the DFT reagents they produced (McKinney, 1985). As the numbers of specimens examined by IFT in Britain is relatively small and there is no evidence to suggest the CDC criteria are unsuitable they are listed here.

> Each smear should be examined for at least five minutes before being considered negative. In the case of lung specimens if 25 or more fluorescent bacteria of the correct morphology are seen the smear is considered positive. If less than 25 bacteria are seen the number is recorded and the smear reported as 'probably positive'. For respiratory tract fluids, which often have only a few bacteria present, five or more per smear is considered positive.

These criteria are not universally applied, for example the presence of five typical bacilli in a lung specimen is considered diagnostic by some workers (E. Dournon, personal communication). The exact number chosen is not important as interpretation is more accurately based on a knowledge of the specificity of the reagents being used and the morphology of any bacteria seen.

ii Specificity and sensitivity

Serological cross-reactions within the family Legionellaceae have been discussed fully in Chapter 6. and are unlikely to cause any diagnostic problems. However cross-reactions of anti-*Legionella* DFT conjugates with bacteria of other genera have been widely reported and may confuse the interpretation of this test. Organisms reported to cross-react include: *Pseudomonas aeruginosa*, *P.alcaligenes*, *P.fluorescens*, *P.maltophila*, *P.putida*, *Bacteroides fragilis*, *Flavobacterium xanthomonas* and *Haemophilus influenzae* (Cherry *et al.*, 1978; Edelstein *et al.*, 1980; Orrison *et al.*, 1983a; Johnson *et al.*, 1985; Tenover *et al.*, 1986). It should be remembered that these cross-reacting strains are the exceptions and represent only a few out of many hundreds examined. In practice the specificity of the CDC conjugates may be in excess of 99% (Edelstein, 1983).

While cross-reactions using the DMRQC rabbit antisera in the IFT have not been reported it is likely that occasional problems of this nature may be encountered. Table 8.1 lists the organisms against which DMRQC reagents have been evaluated. With the exception of the antisera for *L.pneumophila* serogroup 2 and serogroup 3 which show weak fluorescence against several *P.aeruginosa* strains, and *L.hackeliae* serogroup 2 antiserum which shows weak fluorescence against one strain of *Eikenella corrodens*. All other antisera (including the *L.pneumophila* serogroups 1–6 polyvalent serum) appear to be non-reactive against this panel of organisms. A positive result in the IFT is therefore highly indicative of a diagnosis of LD provided that the appropriate controls have been observed.

The sensitivity of the DFT using CDC conjugates has been estimated to be in the range 25–70%, with 60% a probable average (Edelstein, 1984a). Although it has not been thoroughly investigated it is likely that the IFT is at least as sensitive as the DFT. However, because of these poor sensitivities a negative result in the IFT or DFT has no diagnostic value and cannot be taken as evidence that a patient does not have LD.

Table 8.1 Species and numbers of strains tested for cross-reactivity with DMRQC *Legionella* antisera.

Species	Number
Pseudomonas aeruginosa	24*
Pseudomonas fluorescens	4
Pseudomonas putida	3
Flavobacterium breve	1
Flavobacterium sp. Group IIb	1
Flavobacterium sp. Group IIf	1
Bacteroides fragilis	1
Bacteroides melaninogenicus	1
Eikenella corrodens	2
Haemophilus influenzae	2

*Including representatives of each serogroup.

IV Commercial reagents

A number of manufacturers produce antisera for the detection of legionellae but few data have been reported concerning the performance of these reagents (Durham *et al.*, 1984; Tenover *et al.*, 1986). As stated above the specificity of the reagent is of a paramount importance and therefore inexperienced users must be very cautious in the interpretation of their results, particularly where 'contaminated' specimens such as sputa are being examined.

Recently Genetic Systems Corp. produced a FITC-conjugated monoclonal antibody (MAB) directed against a *L.pneumophila* common outer-membrane protein. This reagent has the advantage of reacting with all *L.pneumophila* serogroups and so obviates the need to examine a specimen with multiple antisera preparations. The kit is intended both for the identification of laboratory isolates and demonstration of organisms in clinical specimens. The clinical utility of the kit has been evaluated using a small number of specimens (Edelstein *et al.*, 1985). These authors examined 24 lower respiratory tract specimens from culture-proven cases of LD. All 24 specimens were positive using both the Genetic Systems MAB and CDC polyvalent rabbit antisera. From our experience and that of others (E. Dournon, personal communication) this reagent is very valuable for confirming IFT or DFT results obtained using polyclonal antisera. Unfortunately its utility is limited as it cannot be used successfully on formalin-fixed or histopathologically prepared specimens. Also the fluorescence of legionellae is not as intense as that seen using serogroup-specific antisera and in consequence a false-negative result may sometimes be obtained.

V Detection of antigens in urine and serum

The techniques described above can only be used to establish a diagnosis if sufficient numbers of intact bacteria are present in a clinical specimen. This requirement clearly limits the sensitivity and suitability of such assays. To overcome these limitations several alternative methods have been developed to detect soluble antigens in patient specimens.

Two early studies (Tilton, 1979; Berdal *et al.*, 1979) demonstrated the potential use of enzyme-linked immunosorbent assays (ELISA) to detect antigens in easily obtainable patient specimens (e.g. urine and sputum). Although other workers have developed several assays, including haemagglutination (Mangiafico *et al.*, 1981) and coagglutination (Tang *et al.*, 1982), the majority of published data have been presented in a series of papers by Kohler and colleagues. The first of these (Kohler *et al.*, 1981) reported the use of a solid-phase radioimmunoassay (SPRIA) to detect *Legionella* antigen in urine. Although in this initial study only a small number of specimens from LD patients were examined, the specificity of the assay was evaluated using a large series of over 240 specimens from appropriate control populations and found to be almost 100%. A single false-positive result was obtained using a specimen from a patient who had a nosocomially acquired pneumonia which was probably caused by *Staphylococcus aureus* and *Proteus mirabilis*. Later the sensitivity of the SPRIA was investigated by studying urines collected prospectively from patients during an outbreak of LD (Kohler *et al.*, 1982). Here specimens from 37 LD patients were examined. Antigen was detected in specimens from 90% of those in whom the diagnosis was established by culture or by positive results in both DFT and antibody estimations. However, the estimated sensitivity fell to 50% where the diagnosis was established only by DFT or serology. One of the most significant results in these studies was the demonstration that antigen is detectable in most patients 1–3 days after onset of symptoms and may persist for many weeks (Kohler *et al.*, 1984).

Clearly the routine use of such an assay would allow the diagnosis in most patients with LD to be established early enough to influence clinical management. However, the SPRIA can only be used in a limited number of laboratories equipped to handle radio-isotopes. In consequence the SPRIA was adapted first as an ELISA (Sathapatayavongs *et al.*, 1982) and then as a latex agglutination assay (Sathapatayavongs *et al.*, 1983). The ELISA had essentially the same specificity and sensitivity as the SPRIA but required a longer incubation time. The latex agglutination, although simple and rapid to perform, was not as sensitive or specific as either of the other two assays.

The major drawback with both the SPRIA and ELISA is the preparation of suitable rabbit antisera for antigen capture. Kohler and Sathapatayavongs (1983) report that of six rabbits inoculated with an autoclaved preparation of *L.pneumophila* serogroup 1, satisfactory antiserum was obtained from only one. Furthermore the sensitivity of the SPRIA was 16–32-fold increased using antisera taken more than 9 months post-immunization compared to those taken 6–7 weeks post-immunization. The difficulty of reproducing large quantities of suitable antiserum has hampered workers' attempts to establish this type of ELISA as a routine laboratory assay.

An obvious solution to the problem of rabbit antiserum production is to employ MABs and work is in progress in this laboratory towards such an assay. The difficulty with this approach is the initial production of a MAB which will 'capture' antigen from any isolate of *L.pneumophila* serogroup 1 or preferably any *L.pneumophila* serogroup. The feasibility of this approach has been demonstrated by Bibb and colleagues (1984) who examined specimens from three culture-proven cases of LD using both a conventional direct ELISA and an indirect ELISA which incorporated MABs and rabbit antiserum. They were able to detect antigen in urine specimens and also in serum specimens. Although urine is usually a simple and convenient specimen to obtain, in some patients (e.g. those with renal failure), serum is more suitable, however previous studies had failed to demonstrate antigen in serum (Kohler and Sathapatayavongs, 1983). In addition, by varying the MAB used in the assay, the authors were able to determine the subgroup of the *L.pneumophila* strain from which the antigen was derived without its prior isolation. Such an approach could have considerable practical value in the investigation of outbreaks.

A Laboratory Manual for *Legionella*
Edited by T. G. Harrison and A. G. Taylor

CHAPTER 9

The Diagnosis of Legionnaires' Disease by Estimation of Antibody Levels

T. G. Harrison and A. G. Taylor

I General introduction

Despite almost ten years of research into alternative methodologies the estimation of antibody levels is still the most commonly employed method for the diagnosis for Legionnaires' disease. Historically, the indirect immunofluorescent antibody test (IFAT) was the assay used to demonstrate antibodies against the 'Legionnaires' disease bacillus' and thus to implicate it as the causative agent of the 1976 Philadelphia outbreak (McDade *et al.*, 1977).

This assay was further developed in the USA by the Centers for Disease Control (CDC) and in the UK by DMRQC. Both these groups put considerable effort into the standardization of their respective IFATs and made the necessary reagents generally available to diagnostic laboratories. It is largely because of these two factors that the IFAT has been so widely used and remains the serological 'method of choice' despite the subsequent development of alternative assays.

A wide range of other serological assays has been reported in the literature, each having its own advantages and disadvantages when compared to the IFAT. Enzyme-linked immunoassays (ELISAs) are extensively used for the diagnosis of viral infections because they are sensitive, can have an objectively determined (spectrophotometric) endpoint, allow automation and enable soluble antigens to be immobilized. In an attempt to implement some of these advantages in the diagnosis of LD several ELISAs have been developed and at least one is now commercially available.

Various approaches to the preparation of antigens have been tried for these ELISAs. For example killed bacteria have been treated in various ways to extract soluble antigens by EDTA extraction (Wreghitt *et al.*, 1982), PBS extraction (Farshy *et al.*, 1978) and bacterial sonication (Elder *et al.*, 1983). Alternatively killed whole organisms have been immobilized on the solid phase (Herbrink *et al.*, 1983; Barka *et al.*, 1986). As none of these antigens has been purified they compromise a complex mixture of bacterial proteins and lipopolysaccharides (LPS) and the use of these crude antigen preparations may be responsible, in part, for the disappointing results obtained. Where increased sensitivity (compared to the IFAT) has been demonstrated there has been a corresponding decrease in specificity (Elder *et al.*, 1983) and alternatively where specificity was found to be good the sensitivity was poor (Farshy *et al.*, 1978). Thus to date the advantages of ELISAs do not outweigh their disadvantages and consequently they are not widely used.

In addition to the IFAT and ELISA, assays which are technically simple to perform have been developed for use in routine laboratories. Farshy and colleagues (1978) adapted a microagglutination test (MAT) for the detection of brucellosis (Gaultney *et al.*, 1971) for the diagnosis of LD. Although this is a simple assay it has not been fully evaluated for sensitivity and specificity. However, this test has found wide application in various forms and is frequently used for serosurveys (Helms *et al.*, 1980; Collins *et al.*, 1982; Lind *et al.*, 1983; Temperanza *et al.*, 1986).

Compared to the IFAT these MATs are generally less sensitive (Farshy *et al.*, 1978). One modification of the MAT which appears to overcome this, is the inclusion of a centrifugation step (Harrison & Taylor, 1982b). In addition to increasing test sensitivity this has the added advantage of reducing the test time from over 18 hours to about 30 minutes. This assay, the rapid microagglutination test (RMAT), is described in detail below.

Direct agglutination of bacteria is also employed in another test. Kleger and Hartwig (1979) modified the plasma reagin test for syphilis to provide a *L.pneumophila* card agglutination test. The small number of specimens examined do not allow this assay to be evaluated.

Agglutination of erythrocytes has been utilized in two assays. One, an immune adherence haemagglutination assay (Lennette *et al.*, 1979), relies on the complement-mediated agglutination of erythrocytes with antibody-coated bacteria, while the second is a direct agglutination of erythrocytes sensitized with soluble *L.pneumophila* antigens (Edson *et al.*, 1979). The limited evaluations of these assays reported in the literature do not show any advantages over the much simpler bacterial agglutination tests described previously. The same can also be said of an immunodiffusion test described by Soriano and colleagues (1982) and a counter immunoelectrophoresis (CIE) assay developed by Holliday (1980).

Many of the assays outlined above have been used for the diagnosis of *Legionella* infections caused not only by *L.pneumophila* serogroup 1, but also by other *L.pneumophila* serogroups and *Legionella* species (Helms *et al.*, 1980; Herbrink *et al.*, 1983; Lind *et al.*, 1983). Although in some instances attempts have been made to evaluate the sensitivities of these assays the limited availability of specimens from appropriate culture-proven cases has made this impossible to date. The remainder of this chapter is therefore primarily concerned with the diagnosis of *L.pneumophila* serogroup 1 infections.

II The antibody response

Results obtained in studies using antibody class-specific conjugates in both the IFAT and ELISA show that the antibodies detected in a serum specimen may be of immunoglobulin-M, -G or -A classes. These may be present either singly or in combination (Wilkinson *et al.*, 1979a). Usually both IgM and IgG can be detected in patient specimens during the course of their illness, but in a significant proportion of patients no IgG is detectable. For example, in a study using the IFAT and RMAT, 29/129 (22%) patients with serologically proven LD had no detectable IgG although appropriately timed serum specimens were examined (Harrison, 1984). Conversely, a few patients, 4/129 (3%), clearly produced IgG but IgM was not detectable.

The quantity of IgA present in sera from LD patients has not been studied in detail. However, the limited evidence available suggests that it is detectable in most specimens and in some may be the major immunoglobulin class present (Wilkinson *et al.*, 1979a).

It is clear from the above data that the serological diagnosis of LD can be best effected using an assay which detects all three major classes of immunoglobulins, and for this reason in the IFAT described below an anti-human whole-immunoglobulin FITC-conjugate is used.

The stage during the course of illness at which specific antibodies become detectable varies from one individual to another and is also dependent upon the assay in use. An early study of 22 patients found that in the IFAT antibody was not detectable until eight days or more after onset of symptoms (Nagington *et al.*, 1979b). However, this was only a small series of patients and in addition it is difficult to determine accurately the date of onset of illness particularly in the case of community-acquired infections.

In the clinical context the delay between hospital admission and diagnosis is critical and this can be accurately established. In a recent study of 119 patients with bacteriologically proven LD (Harrison *et al.*, 1987) 40% of specimens taken within a week of hospital admission had significant levels of antibody detectable by IFAT, and 45% had levels detectable by RMAT. Although the antibody levels in the majority of these early specimens were below the diagnostic threshold, a significant proportion showed diagnostic titres (16% and 23% for the IFAT and RMAT respectively). In this study approximately 30% of patients acquired their illness nosocomially and for these cases exact dates of onset of symptoms were known. The average period of time between onset of illness and the earliest detection of antibodies was not significantly different in the nosocomial group from the time between hospital admission and detection of antibodies in patients with community-acquired LD. These findings suggest that antibodies are often present soon after the onset of symptoms.

This study confirms the observation reported by other workers (Dournon *et al.*, 1983) that most immunocompromised patients show a normal antibody response. In the study approximately 50% of the 119 patients were immunocompromised and yet the majority showed no significant differences in antibody titres when compared with those in the non-immunocompromised group. There were, however, a small number of severely immunocompromised patients who failed to produce detectable antibody.

There is some evidence that IgM may reach detectable levels before IgG (Nagington *et al.*, 1979b; Zimmerman *et al.*, 1982; Elder *et al.*, 1983) but this has only been shown for small numbers of patients. In the authors' experience the delay in the appearance of IgG after IgM is generally short in the majority of patients but may in some cases be marked. This delayed IgG response may explain the observation that occasionally an early serum specimen is positive in an agglutination assay but negative in the IFAT (Harrison *et al.*, 1987; Temperanza *et al.*, 1986). Although the question whether IgM production significantly precedes that of IgG remains to be resolved it is clear that the separate estimation of IgG and IgM has no advantage over the estimation of total Ig in the diagnosis of acute infection. The frequently held belief that the presence of IgM indicates recent infection is mistaken in the case of LD and, indeed, other Gram-negative infections, because IgM often persists at high levels for a prolonged period of time after infection (Table 9.1).

Table 9.1 Examples of the variation in the antibody response seen in LD patients.

Patient no.	Days after onset	Days after admission	Antibody titres: IFAT				Comments
			Ig[1]	IgM	IgG	RMAT	
1	−61[2]	–	<16	<16	<16	NT[3]	IgM only,
(45,F)[4]	6	–	64	32	<16	64	falling quickly
	11	–	256	NT	NT	NT	(positive soon
	20	–	256	NT	NT	NT	after onset)
	28	–	32	32	<16	64	
	37	–	16	16	<16	32	
	88	–	<16	<16	<16	<8	
2	9	0	<16	<16	<16	<8	Mainly IgG
(48,F)	15	6	256	16	256	16	response
	23	14	512	64	512	32	
	48	39	128	16	128	<8	
3	9	1	<16	<16	<16	<8	Initially IgM
(49,M)	17	9	≥512	≥512	16	512	response, IgG
	22	14	512	512	32	256	following and
	37	29	512	256	128	128	remaining high
	72	64	256	128	256	128	
4	4	0	<16	<16	<16	<8	As above
(28,M)	6	2	<16	<16	<16	<8	
	11	7	512	512	32	128	
	15	11	1024	1024	512	2048	
	29	25	1024	512	1024	256	
	32	28	1024	512	512	128	
	56	52	1024	512	1024	128	
	230	226	512	<16	512	32	
5	1	11	16	<16	<16	<8	Mainly IgM
(44,M)	22	12	256	128	<16	128	response, titre
	43	33	2048	1024	<16	256	remaining high
	55	45	2048	512	64	256	
	275	265	128	128	16	32	
6[5]	11	3	64	NT	NT	8	Parallel IgM
(29,M)	13	5	256	NT	NT	128	and IgG. A
	17	9	1024	1024	512	512	diagnostic rise
	66	58	1024	512	512	512	established
	93	85	512	512	512	512	soon after
	99	91	128	128	128	128	admission to
	105	97	128	64	128	128	hospital

(*Table 9.1 continued overleaf*)

Table 9.1 contd. Examples of the variation in the antibody response seen in LD patients.

Patient no.	Days after onset	Days after admission	Antibody titres: IFAT Ig[1]	IgM	IgG	RMAT	Comments
7[5]		NK[6]	< 16	< 16	< 16	< 16	Separate estim-
(58,F)	7	NK	16	< 16	< 16	32	ation of IgM
	10	NK	256	128	128	128	and IgG would
	> 10	NK	256	64	128	64	miss the response seen on day 7
8	7	NK	16	NT	NT	NT	Persistently
	13	NK	⩾512	NT	NT	NT	elevated levels
	18	NK	⩾512	512	128	512	of both IgM
	25	NK	⩾512	512	128	1024	and IgG, gradu-
	42	NK	⩾512	512	256	512	ally falling over
	72	NK	256	256	128	128	three years
	181	NK	128	128	128	64	
	287	NK	128	64	64	64	
	529	NK	128	64	64	32	
	918	NK	64	32	32	32	
9[5]	4[2]	–	16	NT	NT	16	Poor antibody
(18,M)	16	–	32	NT	NT	128	response but
	26	–	64	NT	NT	64	positive soon
	32	–	64	NT	NT	32	after onset of
	36	–	64	NT	NT	32	symptoms
	44	–	64	NT	NT	16	
	51	–	16	NT	NT	16	

1. Rabbit anti-whole human immunoglobulin conjugate.
2. Legionnaires' disease nosocomially acquired.
3. NT – not tested, serum specimen not available for testing.
4. (45,F) – a 45-year-old female.
5. Diagnosis proven by culture of *L.pneumophila* serogroup 1.
6. NK – not known, data not available.

Table 9.1 also illustrates the range of antibody responses that may be encountered. Typically the rise in IgM and IgG is rapid and broadly parallel. Titres remain elevated for a variable period of time; the IgM may then fall away leaving high IgG levels, both IgM and IgG may fall away together or, as is often seen, both IgG and IgM remain elevated for an extended period. The reasons for these prolonged elevated antibody levels has not been investigated in detail but may, in part, be due to the T-cell independent and highly immunogenic nature of LPS antigens. Recent studies using immunoblotting techniques show that the major antibody response in LD patients appears to be directed against the lipopolysaccharide components of the *Legionella* cell wall, although antibodies against a diverse range of protein components are also present (Brown *et al.*, 1986; Sampson *et al.*, 1986; Samuel *et al.*, 1987). A second reason for the prolonged antibody response may be the continued presence of *Legionella* antigens in LD patients. This is well illustrated in a study by Kohler and colleagues (1984). They found that 10/23 patients were still excreting urinary antigen for at least 42 days after onset of symptoms despite appropriate treatment and apparent clinical recovery. In one patient the last urine specimen in which antigen was detected was obtained 326 days after onset.

The range of specificities of antibody produced by patients infected by *L.pneumophila* serogroup 1 is wide. In some instances the antibodies react only with organisms of the same serogroup as the infecting strain, but in the majority the antibodies have multiple specificities reacting with strains of several or all serogroups. Inter-serogroup cross-reactions between serogroups 1 and 6, 3 and 6, and 4 and 5 are particularly common (Wilkinson *et al.*, 1983; Harrison, 1984). The titres of sera from culture-proven cases examined in DMRQC, and elsewhere, are as great, or usually greater, against strains of the homologous serogroups as against strains of heterologous serogroups. The large numbers of sera from culture-proven *L.pneumophila* Sgp1 infections which have been examined to date, suggest that antibody assays can be used to identify the serogroup of the infecting strain in such cases. However, the few data available from culture-proven cases caused by serogroups other than serogroup 1 do not allow the same assumption to be made regarding antibody tests for these infections.

Patients with LD may also produce antibodies which will react against antigens common to both *Legionella* and other bacteria. This has been shown by several groups of workers (Wilkinson *et al.*, 1979a; Collins *et al.*, 1983, 1984) and has undoubtedly confused the interpretation of some serological assays used to diagnose LD (Tsai & Fraser, 1978). The nature of these antigens has not been determined but they do not appear to be detected using the IFAT or RMAT described in this chapter.

III Specimen collection and storage

Venous blood is collected into a suitable sterile container, allowed to clot and the serum separated by centrifugation. Separated plasma is also satisfactory for examination. Post-mortem blood specimens can be used but as these are usually grossly contaminated the results obtained using such specimens must be interpreted with care.

A serum specimen should be collected from patients and tested as soon after the onset of symptoms as possible, since in a significant proportion of cases antibodies are detectable on admission to hospital. The likelihood of demonstrating rising antibody levels is greatest if specimens are taken repeatedly (ideally not more than three days apart) for up to three weeks after onset of symptoms. It is very unusual for a patient who will subsequently seroconvert not to have detectable antibody by this time.

Specimens can be satisfactorily stored at 4 °C for at least one week but if testing is to be seriously delayed preservative should be added (e.g. 0.08% NaN_3) or alternatively the specimen should be stored frozen at or below −20 °C.

Careful storage of specimens is important as they may be needed for parallel testing with other sera from the same patient. Microbial contamination may interfere with assay results by giving 'false-positive' agglutination or by degradation of serum antibodies. Specimens stored at a constant temperature of below −20 °C appear to maintain their antibody titres over several years (both IgG and IgM). Repeated freezing and thawing may reduce the antibody titre and should be avoided.

IV The indirect immunofluorescent antibody test (IFAT)

A Introduction

As discussed above the IFAT is usually considered the serological reference test against which other assays are compared. However, there are many factors which influence the results obtained by this technique and these include: the antigen (strain, growth conditions, killing and fixation methods, and diluent), the conjugate (class specificity and titre), microscopy (light source, epi or trans illumination, filters and magnification), the definition of the assay endpoint, and the subjective nature of the observation. The careful use of internal standards can overcome variation introduced by instrumentation and observer bias. The choice of conjugate can entirely alter the nature of the test by restricting the immunoglobulin classes detectable but can be easily standardized. However, the choice of antigen is the most contentious and variable component and so will be discussed in more detail here.

The IFAT first described by McDade and colleagues (1977) used a live preparation of infected chick embryo yolk-sacs as the antigen. These authors had little choice of antigen as at the time, apart from guinea-pigs, this was the only system in which the organism was known to grow. The yolk-sac method of antigen production was adopted in this laboratory when work started on *L.pneumophila* in 1978 but inactivation of the bacteria with formalin avoided the need to use live material (Taylor *et al.*, 1979). The wide demand for reagents in the USA required the CDC to produce large quantities of killed antigen. They chose to grow the organism on, the then recently described, appropriate solid culture media and to kill the bacteria with diethyl ether (Wilkinson *et al.*, 1979b).

An early study found that false-positive results were obtained using the ether-killed antigen which were not seen using the formolized yolk-sac antigens (FYSA) (Taylor *et al.*, 1979). Subsequently the CDC encountered other problems and abandoned ether-killing in favour of heat-killing. This solid media grown, heat-killed antigen has proved to be satisfactory and is the antigen preparation method which the CDC still recommend.

Despite reports to the contrary (Edelstein, 1984a) there has been no published comparison of the DMRQC and CDC *L.pneumophila* Sgp1 antigens so neither can be said to be superior. Studies have been published which compared antigens prepared by heat-killing with those prepared by formalin-killing and no significant differences were found when each assay was appropriately standardized (Wilkinson & Brake, 1982). However, the two antigens differ not only in killing methods, but in growth conditions (eggs or solid media) and the strain of *L.pneumophila* serogroup 1 used and these factors were not considered. It is clear however from the published data that in areas where FYSA is used routinely (UK and France), the antibody levels encountered in non-LD populations are sufficiently low to allow diagnostic significance to be attached to even moderately elevated antibody levels (Macrae & Lewis, 1977; Harrison & Taylor, 1982a; Dournon *et al.*, 1983). Generally this does not seem to be true in areas where the CDC antigen is used. Whether this is due to antigen or demographic differences has not been established.

B Preparation of formolized yolk-sac antigens (FYSA)

The *Legionella* strain to be used is inoculated on to BCYE and incubated at 37 °C for 48–72 hours. The growth is washed off into sterile distilled water and the concentration of bacteria adjusted to approximately 10^8 organisms/ml. The suspension (0.2 ml) is inoculated into the yolk-sacs of six- or seven-day-old embryonated hens' eggs.

Alternatively, the freeze-dried contents of a NCTC culture ampoule are reconstituted with 1 ml of nutrient broth and inoculated (0.2 ml) directly into the yolk-sacs of embryonated hens' eggs (0.2 ml contains approximately 2×10^7 viable organisms).

The inoculated eggs are incubated at 33 °C and 'candled' daily to check their viability. The yolk-sacs of embryos dying between two and seven days after inoculation are harvested and pooled; eggs with live embryos are discarded. These preparations are then homogenized in an equal volume of 2% formalin in distilled water to prepare the stock antigen. Two or three eggs from each batch are homogenized in an equal volume of sterile distilled water to prepare seed antigen. To ensure the seed antigen contains adequate numbers of viable organisms a BCYE plate is inoculated with the seed and incubated at 37 °C. The stock-formolized antigen is incubated overnight at 37 °C and a sample inoculated on to BCYE to ensure the organisms in this preparation have been killed.

The 'passage' history and storage conditions are important factors in the choice of strains used to prepare antigens. Clinical isolates which have been subcultured on solid media only a few times will usually kill eggs rapidly and antigens made from the yolk-sac will be suitable for use at high dilutions. However, some strains require two or three yolk-sac passages before producing a lethal infection within seven days. In such cases 0.2 ml of live seed antigen from the previous passage is used as the inoculum. Strains may be encountered which will not kill eggs even after multiple yolk-sac passage.

To establish the working strength of each antigen batch, dilutions are made in Dulbecco A phosphate buffered saline pH 7.2 (PBS) and examined by the IFAT using rabbit antiserum raised against the homologous strain. The antigen dilution in which about 200 morphologically recognizable organisms per microscope field (×10 eyepiece, ×63 water objective) are seen is selected as the working dilution. Most antigens are suitable for use at a dilution between 1/25 and 1/80 (each egg harvested provides antigen for coating approximately 20 000 wells in the IFAT).

The stock antigens are stored undiluted at 4 °C and are stable for at least five years. Before use the stock antigens are diluted to working strength in PBS containing 0.08% sodium azide and passed through a fine cotton gauze to remove any large particulate matter. Antigens at working dilutions are stable for at least one year if stored at 4 °C.

DMRQC antigen

The antigen supplied by DMRQC for use in the standard IFAT is FYSA and has, until recently, been made from the Pontiac-1 strain (NCTC 11191) of *L.pneumophila* serogroup 1. The antigen issued from May 1986 is made from a very similar strain of *L.pneumophila* serogroup 1 isolated in the UK from a patient with LD. These two strains are indistinguishable by monoclonal antibody subtyping, and extensive studies have shown them to give similar results in the IFAT.

C The IFAT using FYSA for the detection of antibodies against *L.pneumophila* serogroup 1

i Preparation of slides

Antigen (5 μl) at the working dilution is applied to each 3 mm well of a PTFE-coated microscope slide. The spots are then air-dried for 20 minutes at 37 °C and fixed for 15 minutes at room temperature in acetone. If 3 mm wells are not used the volume of antigen must be adjusted. However, there is no advantage in using larger wells and excessive quantities of antigen will then be required.

ii Storage of slides

Antigen-coated slides are best prepared on the day of use but can be stored satisfactorily overnight at room temperature for use the next day. In laboratories which use only a small number of slides it is sometimes the practice to prepare a batch of slides and store them at −20 °C in a sealed container. This is a satisfactory procedure provided that when examined the number, morphology and staining intensity of organisms seen in the FYSA tested with the human reference serum is typical. Slides frozen in this way should be pre-incubated at 37 °C for 5 minutes before use to avoid condensation forming.

iii Test procedures

Serum dilutions

These are prepared in PBS using doubling dilutions starting at 1 in 16. If a large number of serum specimens are being examined it is often most economical to examine them at screening dilutions of 1 in 16 and 1 in 32 only, and then to titrate any serum specimens found to be positive. If the number of specimens is small it is preferable to titrate them all from 1 in 16 to 1 in 512 from the outset.

To each FYSA spot is added 10 μl of serum dilution (again the volume must be adjusted if the well size is not 3 mm). The slides are incubated in a wet box at 37 °C for 30 minutes. If a large number of slides are being processed on one day it is best to divide them into several batches. If this is not done the time interval between adding serum to the first and last slides becomes a significant proportion of the 30 minutes' incubation and may affect the results.

Washing

The slides are rinsed with PBS and then washed for 15 minutes with two changes of PBS. The slides are briefly rinsed with distilled water, gently blotted and dried at 37 °C (10 minutes).

FITC conjugate

To each well is added 5 μl of conjugate dilution, the slides are then incubated in a wet box at 37 °C for 30 minutes and washed as above.

After drying, the slides should be viewed immediately or stored in the dark if this is not possible. The slides should be examined on the same day as they are prepared, after this time the fluorescence tends to fade and the rate of fading may vary from serum specimen to serum specimen.

Microscopy

The slides are mounted in glycerol mounting media or viewed unmounted with a water immersion objective. Where slides are mounted care must be taken to avoid trapping air between the slide and coverslip as positive fluorescence may then be overlooked.

Preparation of glycerol mounting media:

Buffered saline* pH 8.5	1 part
Glycerol, neutral	9 parts

Prepare 0.067 M K_2HPO_4 by dissolving 1.61 g in 100 ml 0.85% NaCl. Prepare 0.067 M KH_2PO_4 by dissolving 0.907 g in 100 ml 0.85% NaCl.

The slides are examined by epi-illumination using ×10 eyepieces and a ×100 oil objective or a ×63 water objective. Microscopes using transmitted light do not generally have adequate sensitivity while objectives of less than ×63 do not give sufficient magnification to allow the morphology of the organism to be seen clearly and hence non-specific fluorescence may not be recognized as such. We use a Zeiss microscope equipped with a HB050 mercury vapour lamp and the following interference filters: excitation filters – LP 455 nm and KP 490 nm, dichromic mirror – FT 510 nm, barrier filters – BP 520–560 nm and BP 590 nm.

The fluorescence is scored + + +, + +, +, ±, – and must be associated with morphologically recognizable legionellae. Fluorescence scored as '+' is defined as being of sufficient intensity for most of the organism present in the antigen to be seen clearly but not brightly. A small number of weakly fluorescent bacteria would be considered '±' and no organisms visible '–' (Plate IVm,n).

iv DMRQC human reference serum

The correct use of the positive control (human reference serum) allows IFAT results to be standardized both within and between laboratories (Taylor & Harrison, 1983). This serum must be titrated (1 in 16 to 1 in 512) on every occasion that the assay is performed and test results calibrated accordingly. The positive control serum should give '+' fluorescence at the dilution stated on the label (usually 1 in 128). Before other test results are determined the positive control specimen should be examined. Strong fluorescence (+ + +) of all bacteria should be seen in the 1 in 16 dilution. As the dilution increases the degree of fluorescence and number of organisms seen should decline. The intensity of fluorescence and number of bacteria visible at the dilution stated to give '+' fluorescence (1 in 128) is noted. If this does not correspond to '+' fluorescence as defined above (iii) the assay is being performed under suboptimal conditions. If '+' fluorescence is seen either one dilution below (1 in 64) or above (1 in 256) this is considered as acceptable experimental error, but if it is outside this range all test results determined on that occasion are void.

* Ten parts 0.067 M K_2HPO_4 in 0.85% NaCl plus one part 0.067 M KH_2PO_4 in 0.85% NaCl.

In addition to the positive control serum, a serum specimen from a healthy individual (negative control) should be included each time the assay is performed. No fluorescence should be seen at a dilution of 1 in 16.

To help conserve the DMRQC human reference serum, laboratories which perform the IFAT frequently may prefer to calibrate their own 'in house' positive control serum. The specimen chosen should have a similar titre to the DMRQC serum (i.e. 64 or 128) and should be shown to give a consistent titre, both within and between assay runs. Both the DMRQC and 'in house' controls should be compared regularly to ensure continued calibration. The use of very high titred sera (>128) as reference preparations is not advisable as these will only reveal gross changes in test performance.

v Determining the antibody titre of test specimens

The titre of a test specimen is the reciprocal of the serum dilution giving the same degree of fluorescence as seen with the positive control serum at its stated '+' dilution (1 in 128). In most instances this corresponds to '+' fluorescence as defined in (iii) above. If sera are found to be positive at screening dilutions they must be titrated to determine the endpoint, as this cannot be estimated by the intensity of fluorescence seen at these low dilutions. Also if an unexpected number of positive sera are found on the same slide, these results should be confirmed before being reported.

vi Conjugate titration and storage

An anti-human whole-immunoglobulin FITC conjugate must be used. Both Wellcome Diagnostics MFO1 (sheep) and Institut Pasteur 74511 (goat) conjugates have been found to be satisfactory. Other manufacturers produce similar products, but these have not been examined in this laboratory.

Each batch of FITC conjugate must be titrated to determine its optimum dilution for use. A doubling dilution series is made from 1 in 10 to 1 in 160 and used in the IFAT. The highest dilution of conjugate giving the appropriate titre with the human reference serum is chosen for use.

After reconstitution the conjugate should be divided into aliquots and stored at −40 °C until required. When needed the conjugate should be diluted to working strength in PBS containing 0.08% sodium azide as preservative. In our experience diluted conjugates stored at 4 °C are suitable for use for up to two weeks.

vii Established diagnostic criteria

The criteria for the serological diagnosis of Legionnaires' disease by the IFAT using *L.pneumophila* Sgp1 FYSA is a four-fold or greater rise in titre to ⩾64. A titre of ⩾128 in a single serum or serial specimens from a patient with a relevant clinical history is accepted as evidence of a presumptive case.

Table 9.2 IFAT titres in sera from 317 patients with infections of known aetiology.

Aetiological agent	Evidence of infection	No. of patients	No. of serum specimens	Titre (no. of serum specimens)
Mycoplasma pneumoniae	Isolation of *M. pneumoniae*	21	42	< 16 (all)
	IgM positive in IFAT[1] or MHAT[2]	10	18	< 16 (all)
	4-fold rise in CFT[3] to ⩾64	8	16	< 16 (all)
	single/standing CFT ⩾64	20	36	< 16 (all)
Chlamydia psittaci (51 patients)	Outbreak[4] (agent isolated from duck carcasses) with CFT ⩾64	7	7	< 16 (all)
	Outbreak (not proven by isolation) with CFT ⩾64	10	10	< 16 (all)
	4-fold rise to ⩾32 or single titre ⩾64 in CFT	34	47	< 16 (all)
Coxiella burnetti (11 patients)	Rise in CFT from <8 to ⩾256	11	22	< 16 (all)
Mycobacterium tuberculosis (33 patients)	Isolation of *M. tuberculosis* from the sputa of patients with active TB	33	33	< 16 (32) 16 (1)
Influenza A (6 patients)	4-fold rise in CFT to ⩾320	6	12	< 16 (all)
Adenovirus (18 patients)	Outbreak with isolation of Adenovirus type 4 (hospitalized patients)[5]	16	32	< 16 (all)
	CFT titre ⩾64	2	3	< 16 (all)

Leptospira interrogans (49 patients)[6]	CFT titre ≥ 80 or MAT[7] titre ≥ 100	49	56	< 16 (all)
Salmonella typhi (55 patients)	Isolation of *S. typhi* from blood or faeces, with positive serology[8]	52	52	< 16 (all)
	Isolation of *S. typhi* only	3	3	< 16 (all)
Bacteroides fragilis (13 patients)	CIEP[9] +ve for specific antibody	9	14	< 16 (all)
	Isolation of *B. fragilis*	4	4	< 16 (all)
Pseudomonas aeruginosa (22 patients)	Isolation of *P. aeruginosa* with positive serology[10]	22	28	< 16 (26) 16 (2)

1. IFAT = Indirect immunofluorescent antibody test (Sillis & Andrews, 1978).
2. MHAT = Microhaemagglutination test (E. O. Caul, unpublished method).
3. CFT = Complement fixation test (Bradstreet & Taylor, 1962).
4. Outbreak of psittacosis in a duck processing factory.
5. Outbreak of Adenovirus infection in military recruits.
6. *Leptospira* serotypes: hebdomadis (14), icterohaemorrhagiae (22), canicola (2), unknown (11).
7. MAT = Microscopic agglutination test (Turner, 1968).
8. Either *S. typhi* Vi agglutination positive (≥ 10) or *S. typhi* Vi IFAT positive (≥ 32) (Doshi & Taylor, 1984).
9. CIEP = Counterimmunoelectrophoresis.
10. Indirect haemagglutination test positive (≥ 160) (H. Todd, unpublished method).

D Interpretation of test results

i Applications

The IFAT described above was initially developed to aid the differential diagnosis of Legionnaires' disease from other pneumonias. It has become clear that *L.pneumophila* infections may be manifested as a wide spectrum of clinical conditions in addition to atypical pneumonia (see Chapter 2). The IFAT is sometimes used to help establish the diagnosis in patients suffering from some of these non-pneumonic forms. However, results obtained with serum specimens from such patients must be interpreted with caution. The test specificity has been evaluated with regard to atypical pneumonias and the range of organisms causing them, and may be different for infections where other organisms are likely causes.

ii Specificity

The specificity is the most important parameter of an assay for an illness of low prevalence such as LD (see iv below). Table 9.2 shows the results obtained using 433 serum specimens from 317 patients with either non-LD respiratory infections, or other illness reported to interfere with the IFAT. It can be seen that sera from only three patients gave even low positive titres (16). Thus the specificity of the assay with regard to diagnostic results is 100% and with regard to any positive result (titre $\geqslant 16$) is 99%.

Two of the three patients whose sera gave low positive results had *Pseudomonas* infections and in one this was systemic. Titres of 16 or 32 have been seen in several patients with systemic *Pseudomonas* infections (E. Dournon, personal communication) and caution should be exercised when examining sera from such patients. Similarly sera from intravenous drug abusers appear occasionally to give low false-positive results. It should be stressed that in both these groups of patients the levels of antibody detected are low, and although a diagnosis of LD may be initially considered the failure of such patients to seroconvert would prevent misdiagnosis.

iii Sensitivity

The sensitivity of the IFAT has only recently been evaluated because of the difficulty in obtaining adequate numbers of appropriate serum specimens. Sera from 119 patients with bacteriologically proven *L.pneumophila* serogroup 1 infections have now been examined and the test sensitivity found to be 79%. In addition a positive (but not diagnostic) result was seen in the cases of a further 10.0% of patients with such proven infections (Harrison *et al.*, 1987). It should be noted that the patients examined in this study were a selected sample and may not have been representative of 'typical LD patients'. For example, the mortality rate was higher than expected and over 50% of patients were immunocompromised. It therefore follows that the estimated sensitivity may not be an accurate estimate, however if this is so it is likely to be an underestimate rather than an overestimate.

iv Predictive values

The predictive value of a positive test result (probability that a patient with diagnostic serology has LD and that the result is not falsely positive) depends on the prevalence of the illness in the population tested, as well as on the test specificity and sensitivity (Vecchio, 1966). An exception is where the specificity is 100%, as found for the IFAT. In this case the predictive value is also 100% irrespective of the prevalence. To date the authors are unaware of any case where a patient shown to have LD by a rising titre in the IFAT which met the diagnostic criteria, has subsequently been shown not to have LD but to have an infection of other aetiology.

v Single or standing titres

The validity of a diagnosis established on the basis of elevated antibody levels in single specimens has been questioned by several workers (Wilkinson *et al.*, 1983; Edelstein & Meyer, 1984). They argue that high titres of antibody are encountered in subjects free of clinical illness sufficiently frequently to cast doubt on the diagnostic significance of single high titres in patients with clinical illness. Although high levels of antibody often persist in patients who have recovered from LD the prevalence of this illness is very low. In addition, multiple episodes of LD seem to be rare and carriage of the organism has not been documented. Also the levels of antibody found in non-LD populations, using this assay, are very low. Therefore, provided the patient does not have a history of LD or pneumonia of unknown aetiology, and presents with an illness compatible with LD it is unlikely that his serum will have high levels of antibody in the absence of LD. There are, however, a few rare instances where this is known to have occurred (E. Dournon, personal communication) and so attempts should always be made to obtain multiple, appropriately timed, serum specimens from suspected LD patients.

vi Weakly positive results

Studies have shown that any positive result ($\geqslant 16$) in the IFAT is rare in the absence of *Legionella* infection. In uninfected control groups the sera from less than 3% of subjects had detectable antibody (Macrae & Lewis, 1977; Harrison & Taylor, 1982a; Dournon *et al.*, 1983). In infected control groups (Table 9.2) titres of $\geqslant 16$ were found in only 1% of sera. This very high level of specificity means that a titre of 16 or 32 in a patient's serum sample is probably of diagnostic significance. As discussed above, studies have shown that the sera from 40% patients who are subsequently shown to have LD have positive IFAT titres early in the course of their illness.

V The rapid microagglutination test (RMAT)

A Introduction

The rapid microagglutination test (RMAT) was initially developed as a technically simple alternative to the IFAT described above (Harrison & Taylor, 1982b). It was envisaged that small laboratories which did not have the equipment or expertise to devote to a relatively uncommon illness such as LD, could use this assay to screen sera from suspected LD patients. Any serum specimen found to be positive would then be forwarded to a reference laboratory for confirmation.

In practice the range of laboratories using the RMAT and their reasons for doing so are somewhat different. The use of the assay has been limited in some small laboratories by the requirement of a centrifuge capable of accepting microtitre plates, while others prefer to continue to use the IFAT which they have found to fit easily into their routine work. Conversely some larger laboratories use the RMAT to screen specimens because it is semi-automatable and has a short assay time. These laboratories generally use the IFAT to confirm their positive results but in some laboratories the RMAT is now the only assay used.

The IFAT and RMAT have been in parallel use in this laboratory for some years now and the data show that the RMAT has the same sensitivity as the IFAT and almost the same specificity (see below). The choice of which of these two assays a laboratory should use is now more likely to be made on considerations of local conditions (e.g. specimen throughput, equipment availability) rather than any differences in assay performance.

B Preparation of antigens for the RMAT

The *Legionella* strain to be used is inoculated on to BCYE agar and incubated at 37 °C for 48–72 hours. The inoculum strain can either be obtained by reconstituting the freeze-dried contents of an NCTC ampoule with 1 ml of nutrient broth, or by recovery and culture of the strain from beads held in liquid nitrogen (Appendix 2).

Culture plates (six inch diameter) of BCYE are heavily inoculated with bacterial growth and incubated at 37 °C for 72 hours. The growth from each plate is then scraped off into 5 ml PBS pH 7.2 (Dulbecco A) containing 2% formalin, and incubated overnight at 37 °C. A fraction of this killed stock suspension is inoculated on to a BCYE plate to test for sterility.

The stock antigen is diluted in 0.1 M PBS pH 6.4* to give an optical density equivalent to OD_{605} = 3.0. Typically a 1 in 100 dilution of stock antigen has an OD of about 0.6, thus the stock is diluted 1 in 20 to give the desired OD. To this is added an equal volume of 0.1 M PBS pH 6.4 containing 0.005% safranin. The safranin is made up freshly on each occasion as a 2% stock solution in methanol. The mixture is then left at room temperature for one hour with occasional shaking and centrifuged at approximately 4500*g* for 20 minutes to pellet the bacteria. The supernatant is discarded and the pellet resuspended in the same volume of fresh 0.1 M PBS pH 6.4 (thus the final OD is equivalent to 1.5). This antigen is used neat in the RMAT. After the addition of preservative, 0.08% NaN_3 or 0.01% merthiolate, the reagent is stable at 4 °C for at least one year.

To check that each batch of antigen is suitable for use a panel of known positive and negative sera are tested. The stained bacteria should pellet well, leaving a clear slightly pink supernatant above. The endpoint should be easy to read and the panel of sera should give the same titres as when previously examined.

C The RMAT for the detection of antibodies against *L.pneumophila* serogroup 1

i Test procedure

To each well of a V-bottomed microtitre plate 25 μl of 0.1 M PBS pH 6.4 is added. The serum sample (25 μl) is then added to the first well of each row and doubling dilutions are made. Antigen (25 μl) is added to each well and the plates are sealed with clear plastic plate-sealers and shaken (either mechanically for 10 seconds using a Dynatech Microshaker, or manually for 20 seconds). The plates are incubated for 10 minutes at room temperature and then centrifuged in microtitre plate-carriers for 4 minutes at approximately 350 *g* (e.g. 1200 rpm in a Beckman J-6B centrifuge with a JR-3.2 rotor). The plate-sealers are removed and the plates inclined at an angle of 70° and left for 10 minutes before reading.

ii Reading the RMAT

This is achieved most easily using a rack made of opaque perspex mounted at 70° to the horizontal and illuminated from behind. A negative result is seen as a button of stained bacteria which has streaked down the side of the well giving a 'tear-drop' pattern. A positive result is usually seen as a tight button with no streaking. At high serum dilutions strongly positive specimens sometimes fail to

* 0.1 M PBS pH 6.4 is prepared as follows: 8.82 g NaH_2PO_4, 3.76 g Na_2HPO_4, 8.5 g NaCl, made up to 1 litre with distilled water.

give a tight button and the bacteria in the well remain diffuse. The endpoint can then be determined by re-reading after a further 10 minutes. The stained bacteria will then have settled on to the lower side of the well at dilutions greater than the endpoint. The titre of a serum is expressed as the reciprocal of the highest serum dilution (before addition of the antigen) giving a positive result (Plate IV.o).

iii DMRQC human reference serum

The positive control (human reference serum) supplied by DMRQC should be used to standardize results on every occasion. This serum should give an endpoint at the dilution stated on the label, if not the titres of test sera should be adjusted accordingly. If the titre of the positive control differs by more than one two-fold dilution from the stated titre the test results are void. In addition to the positive control serum, a 'normal' human serum should be included in each run and should give a titre of <2.

iv Established diagnostic criteria

The criteria for the serological diagnosis of Legionnaires' disease by the RMAT is a four-fold or greater rise in the titre of paired sera to $\geqslant 16$. A titre of $\geqslant 32$ on a single serum specimen from a patient with a relevant clinical history is considered as evidence of a presumptive case.

D Interpretation of test results

i Applications

The RMAT was developed as a technically simple alternative to the IFAT. It is suitable for use by small laboratories which do not have equipment or expertise to use the IFAT, or where the IFAT does not fit easily into the routine of the laboratory. Also the RMAT can be used in large laboratories where substantial numbers of specimens need to be screened rapidly and with the minimum of effort. The major disadvantage of the method is the requirement for a centrifuge capable of accepting microtitre plate-carriers.

ii Specificity

The specificity is the most important parameter of an assay for an illness of low prevalence such as LD. Table 9.3 shows the results obtained using 320 serum specimens from 248 patients with either non-LD respiratory infections, or other illness reported to interfere with *Legionella* serology. It can be seen that sera from 4/143 (2.8%) patients with non-LD respiratory infections gave positive titres.

None of these met the diagnostic criteria but one specimen did give a titre of 16. Sera from 5/105 (4.8%) patients with non-respiratory illness gave positive titres. One of these results, from a patient with a systemic *Pseudomonas* infection, satisfied the diagnostic criteria. Sera from this patient were also positive in the IFAT. Thus the specificity of a diagnostic result in the RMAT is 99.6% (247/248), and for any positive result 96.4% (239/248). It is noteworthy that most of the false-positive results were obtained using sera from patients whose infections were diagnosed using serological assays. It is possible that these were not false-positive RMAT titres but false-positive *Mycoplasma*, *Chlamydia* or *Leptospira* CFT results.

iii Sensitivity

The sensitivity of the RMAT has recently been evaluated (Harrison *et al.*, 1987). Serum specimens from 115 patients with bacteriologically proven *L.pneumophila* serogroup 1 infections were examined and the test sensitivity found to be 81.7%. In addition a positive (but not diagnostic) result was seen in the cases of a further 7.0% of these patients.

iv Predictive values

From the sensitivity and specificity data given above, the predictive value of a diagnostic result is 67–91% assuming that the prevalence of Legionnaires' disease caused by *L.pneumophila* serogroup 1 is 1–5% of patients with pneumonia.

v RMAT positive / IFAT negative results

The RMAT and IFAT have been used in parallel in this laboratory since early 1982. Sera from approximately 3000 patients have been tested since this time, including about 800 LD patients positive in both the IFAT and RMAT. In 40 cases serum specimens gave positive RMAT titres but were negative in the IFAT. Sera from half of these (about 1% of the total population examined) were only weakly positive (8), and might be expected as normal background levels in control populations (Harrison & Taylor, 1982b).

In eleven of the cases negative in the IFAT a diagnostic rise in titre was seen using the RMAT. The explanations for these discrepancies are not known. Some cases may actually be IFAT false negatives, as has been recently reported in one culture-proven case (Harrison *et al.*, 1987), while others are probably RMAT false-positive results. At present there are not enough clinical data to suggest the real aetiology in these instances. Whatever the reasons it is clear that although such cases do occur, false-positive RMAT results are only rarely encountered.

Table 9.3 RMAT titres in sera from 248 patients with infections of known aetiology.

Aetiological agent	Evidence of infection	No. of patients	No. of serum specimens	Titre (no. of serum specimens)
Mycoplasma pneumoniae (50 patients)	Isolation of *M. pneumoniae*	5	10	< 8 (all)
	IgM positive in IFAT[1] or MHAT[2]	4	8	< 8 (all)
	Four-fold rise in CFT[3] to ≥64	6	12	< 8 (all)
	Single/standing CFT ≥64	35	51	16[a] (1) < 8 (50)
Chlamydia psittaci (32 patients)	Outbreak[4] (agent isolated from duck carcasses) with CFT ≥64	7	7	< 8 (all)
	Outbreak (not proven by isolation) with CFT ≥64	10	10	< 8 (all)
	Four-fold rise to ≥32 or a single titre ≥64 in CFT	15	28	8[b] (3) < 8 (25)
Coxiella burnetti (22 patients)	Rise in CFT from <8 to ≥256	22	33	8[c] (2) < 8 (31)
Mycobacterium tuberculosis (33 patients)	Isolation of *M. tuberculosis* from the sputa of patients with active TB	33	33	< 8 (all)
Influenza A (6 patients)	Four-fold rise in CFT to ≥320	6	12	< 8 (all)
Leptospira interrogans (25 patients)[5]	CFT titre ≥80 or MAT[6] titre ≥100	25	28	16[d] (1) < 8 (27)

Salmonella typhi (54 patients)	Isolation of *S. typhi* from blood or faeces, together with serological evidence[7]	51	51	8^{e} (1)
				< 8 (50)
	Isolation of *S. typhi* only	3	3	< 8 (all)
Bacteroides fragilis (13 patients)	CIEP[8] +ve for specific antibody	9	14	< 8 (all)
	Isolation of *B. fragilis*	4	4	< 8 (all)
Pseudomonas aeruginosa (13 patients)	Isolation of *P. aeruginosa* with serological evidence[9]	13	16	32^{f} (1)
				8 (3)
				< 8 (12)

a 1 patient; single serum (RMAT titre = 16; *M. pneumoniae* CFT = 64).
b 1 patient; two sera (RMAT titre = 8 in both specimens; *C. psittaci* CFT = 32 and 64).
1 patient; single serum (RMAT titre = 8; *C. psittaci* CFT = 64).
c 1 patient; two sera.
d 1 patient; single serum (RMAT titre = 16; *Leptospira* sp. CFT = 320).
e 1 patient; single serum (RMAT titre = 8; ViFAT titre = 512).
f 1 patient; two sera (RMAT titre = 8 in both specimens; *Pseudomonas* IHA titre ⩾2560 in both).
1 patient; single serum (RMAT titre = 8; *Pseudomonas* IHA ⩾2560).
1 patient; single serum (RMAT titre = 32; *Pseudomonas* IHA = 10,000).

1. IFAT = Indirect immunofluorescent antibody test (Sillis & Andrews, 1978).
2. MHAT = Microhaemagglutination test (E. O. Caul, unpublished method).
3. CFT = Complement fixation test (Bradstreet & Taylor, 1962).
4. Outbreak of psittacosis in a duck processing factory.
5. *Leptospira* serotype: hebdomadis (5), icterohaemorrhagiae (11), unknown (9).
6. MAT = Microscopic agglutination test (Turner, 1968).
7. Either *S. typhi* Vi agglutination positive (⩾10) or *S. typhi* Vi IFAT positive (⩾32) (Doshi & Taylor, 1984).
8. CIEP = Counterimmunoelectrophoresis.
9. Indirect haemagglutination test positive (⩾160) (H. Todd, unpublished method).

A Laboratory Manual for *Legionella*
Edited by T. G. Harrison and A. G. Taylor

CHAPTER 10

The Application of Nucleic Acid Probes and Monoclonal Antibodies to the Investigation of *Legionella* Infections

N. A. Saunders and T. G. Harrison

I Nucleic acid probes

A Introduction

In recent years it has become clear that, in some instances, nucleic acid probes offer advantages over conventional methods, such as culture and serology, for the detection and identification of microorganisms. The use of probes has been made possible by advances in the technology of nucleic acid manipulation which allow specific base sequences to be isolated and propagated by employing a variety of well characterized host/vector systems. The attraction of using nucleic acid sequences as probes is their potential specificity. The genes of bacterial species are very diverse and significant differences in base sequences are observed even between strains of a single species. Such variations can be used to advantage in the precise identification of bacteria. However, along with this diversity, conserved nucleotide sequences are present and these can be employed for the detection and identification of larger groupings of organisms.

The preparation of polynucleotide probes can be accomplished by cloning, oligonucleotide synthesis or by direct extraction of the desired nucleic acid sequence from an organism of interest. Table 10.1 lists the forms of nucleic probes which have found useful applications because of their different properties. Both cloned and synthetic probes, which consist of defined base sequences, usually have high specificity for target molecules. Furthermore, it is possible to achieve high levels of purity and variations between different batches can be readily controlled. Probes derived directly from organisms of interest are often essential laboratory tools, for example, when it is necessary to screen cloned sequences. They may also be useful in situations where sequence complexity is an advantage such as in some taxonomic studies. Single-stranded probes are generally to be preferred since they can be used in sandwich-type assays and provide advantages in sensitivity compared with double-stranded sequences (Melton *et al.*, 1984).

Table 10.1 Commonly used types of nucleic acid probes.

Derivation of probe sequence	Type of nucleic acid	Double stranded (ds) or single stranded (ss)
Non-cloned sequences	RNA	ss
	DNA	ds
	Complementary DNA transcripts	ss
Cloned sequences	DNA sequences in λ, cosmid or plasmid vectors	ds
	DNA sequences in M13 vectors	ss
	RNA transcripts of cloned DNA sequences	ss
Synthetic probes	Synthetic oligonucleotides	ss

Various methods are available for the detection of target sequences using probes, however most workers follow a method in which the target DNA, attached to a membrane support consisting of either nitrocellulose or nylon, is hybridized to labelled probe sequences. Following hybridization unbound and weakly bound probes can be removed from the membrane by washing under conditions of the required stringency. High stringency washes are under conditions close to the melting temperature (which varies with percentage G+C) of the perfectly base-matched duplex. Mismatched duplexes, which have a lower melting temperature, can be left intact if this is desirable by using less stringent conditions. Following washing, the labelled probe remaining bound to the filter is detected by an appropriate means.

Many systems for the detection of hybridization between probe and target sequences are now available. Unfortunately, despite recent advances in the technology, the systems in use are still not universally applicable. Radioactive probes are widely employed and target sequences can be detected with high sensitivity but their widespread use is limited by the following constraints:

1. The isotopes which offer the greatest sensitivity have relatively short half-lives.
2. Many routine laboratories are not equipped for the safe handling of radioisotopes.

Alternatives to radiolabelling currently under development have great potential. The most widely used non-radioactive labelling and detection systems employ biotin, which can be introduced into the probe by a variety of means (Langer *et al.*, 1981). The biotin is then detected directly, using enzyme-linked streptavidin, or indirectly. In the indirect system streptavidin is first bound to the biotinylated probe and then detected with enzyme-linked biotin.

Alternatively, monoclonal antibodies (MABs), raised directly against nucleic acids (Huang *et al.*, 1985), can be used in an ELISA. Similarly, MABs directed against a hapten introduced into the probe chemically or enzymatically may be employed. Another promising procedure is based on the ability to link enzyme molecules directly to the probe (Renz & Kurz, 1984) so that the number of steps and time required for detection of the target sequence is greatly reduced. In addition the tendency for non-specific binding of streptavidin or antibody is eliminated.

In summary nucleic acid probes already have considerable practical value in the research laboratory and it seems probable that the technical difficulties preventing their more widespread use will be overcome.

B Probes for the detection and identification of bacteria belonging to the family Legionellaceae

Current serological and biochemical indices are inadequate for the identification of strains of some species of the Legionellaceae which must therefore be classified exclusively on the basis of their DNA homologies (Brenner *et al.*, 1984). Furthermore, culture of legionellae is complex and may require from two to ten days to complete. Nucleic acid probes for the recognition of these organisms are under development because they offer the possibility of rapid detection and identification of all species of *Legionella*.

Grimont and colleagues (1985) have described a probe derived from the total genomic DNA of *L.pneumophila* which is specific for this species. The DNA was digested with a restriction endonuclease (Bam HI) and the fragments separated by electrophoresis in agarose gels or by sucrose density gradient centrifugation. DNA molecules coding for parts of the ribosomal RNA (rRNA) genes were excluded (this was efficiently achieved by employing DNA fragments of <9 kb) because these genes are highly conserved and cross-react with the DNA from other bacterial species. The remaining probe sequences were radiolabelled with ^{32}P by nick-translation and tested by dot blotting against DNA derived from other bacterial species belonging to various families including the Enterobacteriaceae, Pseudomonadaceae and Vibrioaceae. The specificity of the probe was found to be adequate and the sensitivity of detection was 10^4 colony-forming units of *L.pneumophila*. When the probe was tested in colony-filter hybridizations it was found that *L.pneumophila* colonies could be detected when they had reached pin-point size (usually after two days' culture). In similar experiments bacteria of other species and genera gave no detectable signal even after extensive growth. *L.pneumophila*, in samples of bronchial exudate or lung tissue, cultured on BCYE agar could be detected using the probe as soon as colonies became visible (2–3 days) while contaminating colonies of non-*Legionella* species gave no signal.

An advantage of this type of probe is that a large proportion of the cellular DNA of the test organisms is a potential target sequence and the theoretical sensitivity is therefore high. It is also advantageous that vector and host DNA, which can be difficult to remove from cloned probes, is absent. A possible disadvantage is the high sequence complexity of the probe resulting in relatively slow hybridization to the target. Also difficulties arise in the reproducible preparation of the probe.

A probe for all species of the Legionellaceae has been produced (Kohne *et al.*, 1984) and is now available commercially from Gen-Probe Inc, San Diego. This probe is produced by reverse transcription of rRNA from a serogroup 1 strain of *L.pneumophila* using universal primers. Sequences common to other bacterial genera are absorbed out to leave a product which hybridizes strongly with rRNA of the Legionellaceae but only weakly with the rRNA from other species. The probe, labelled by radio-iodination, is added to the bacterium in cell lysing solution. Following hybridization the RNA/DNA duplexes are bound to hydroxyapatite which is then collected by low speed centrifugation and washed. The bound probe is then quantified by gamma counting. The rationale of this approach is that being derived from the highly conserved rRNA gene sequences the probe is able to react with a range of *Legionella* species. A further advantage is that the target rRNA sequences are present in the cell in high numbers.

The Gen-Probe *Legionella* probe has been evaluated for the identification of isolates (Edelstein, 1986; Wilkinson *et al.*, 1986). These studies demonstrate that the probe can distinguish *Legionella* from non-*Legionella* species although in one of the studies 4/10 strains of *L. bozemanii* were initially misidentified. The authors used various methods to evaluate the data. Edelstein calculated the percentage hybridization relative to *L. pneumophila* Philadelphia-1 after subtracting non-specific background counts. When the percentage hybridization was $>20\%$ the isolate was considered to be a *Legionella*. In the study reported by Wilkinson and colleagues 4/10 *L. bozemanii* isolates were misidentified when the data were evaluated as recommended by Gen-Probe. In this method the percentage hybridization is calculated as the proportion of total counts bound, after subtracting non-specific background. The figure recommended by Gen-Probe for positive identification of *Legionella* species is 10% hybridization. When this value was recalculated as being greater than or equal to two standard deviations above the maximum value obtained in testing non-*Legionella*, only one of the previously misidentified *L. bozemanii* strains remained negative. It was suggested that the difficulty in identification of these strains was due to inefficient lysis of stored specimens.

The results of a study using the Gen-Probe *Legionella* probe to examine human respiratory tract samples has also been published (Edelstein *et al.*, 1987). Specimens which had been stored frozen at $-70\,^{\circ}$C, for one to eight years, were examined. These comprised 112 culture-positive and 230 culture-negative specimens which were matched, if possible, for both collection date and specimen source. The culture-negative samples were also negative by immunofluorescence (IF). After testing in the Gen-Probe assay the results were expressed as a ratio of counts per minutes (cpm) in the sample to cpm in a negative control. A ratio of 4.0 was considered positive for legionellae. The specificity of the assay was 99.1% and the sensitivity was 56.3%. It was reasonably suggested that the figure for sensitivity may represent a minimum value due to deterioration of bacterial nucleic acids during specimen storage. The sensitivity of the assay was also dependent upon the type of specimen examined, being higher for lung tissues and endotracheal secretions than for sputa and transtracheal aspirates. It was suggested that this could be due to either the lower numbers of organisms present or the greater loss of titre during prolonged storage of these samples. Specimens in the culture-positive group were re-examined and it was found that in some cases legionellae could not be re-isolated. If these samples were excluded the sensitivity of the test was in the range 70–75%. Twelve of 112 samples positive for *Legionella* contained species other than *L. pneumophila*. When tested by the probe four gave a result considered positive. This result is in agreement with the predicted reduction in sensitivity of the probe assay for non-*L. pneumophila* legionellae, which is to be expected because the probe hybridizes less strongly with these species (Edelstein, 1986; Wilkinson *et al.*, 1986).

Cloned probes have been described which are specific for either *L.pneumophila* or for the Legionellaceae. These were derived from selected regions of the gene coding for a 24 kD major outer membrane protein of *L.pneumophila* (Eisenstein & Engleberg, 1986; Engleberg *et al.*, 1986). In colony-filter hybridization experiments a DNA probe consisting of a large proportion of the 5′ half of the gene hybridized only to *L.pneumophila* strains, whereas a probe consisting of the 3′ half of the gene also hybridized with the other *Legionella* species tested. Homology of the probes with non-legionellae was not detected even under conditions of low stringency. The probes were tested for their ability to detect *L.pneumophila* in mouse lung homogenate (following inoculation of 2×10^8 organisms) and the threshold sensitivity was estimated as $5-10 \times 10^4$ organisms.

Studies in this laboratory have been directed towards the production of cloned nucleic acid probes for the following possible applications.

1. Detection of *L.pneumophila* in clinical and environmental specimens.
2. Identification of cultures and detection of *Legionella* species in clinical and environmental specimens.
3. Detection of inter-strain restriction-fragment length polymorphisms (RFLPs) as the basis of a typing scheme for *L.pneumophila*.

We have prepared a gene library from the Knoxville-1 strain of *L.pneumophila* (Saunders *et al.*, 1988). The DNA fragments were generated by Eco RI endonuclease digestion of the genome and pieces of DNA of 5 – 14 kb in size were cloned into the bacteriophage vector λgtWES.λB. (Tiemeier *et al.*, 1976). The cloned DNA fragments used in the studies described below were derived from this library.

A probe for *L.pneumophila* was selected by testing a panel of cloned sequences from the library against purified DNA from various *L.pneumophila* and non-*L.pneumophila* species. The clone chosen λNS21 contains approximately 10 kb of *L.pneumophila* DNA, which was subcloned into a plasmid vector, pT7-1. Figure 10.1 shows a dot blot of various DNAs hybridized with RNA transcripts of the resulting plasmids, which were labelled by incorporation of ^{32}P-labelled UTP. The blots were washed under conditions of high stringency. All of the *L.pneumophila* strains examined, which were of various serogroups, hybridized to the probes. No signal was observed with non-legionellae although on some blots weak hybridization with other *Legionella* species was detected.

It is intended to use these probes to develop a rapid assay system for *L.pneumophila* in clinical and environmental samples and to confirm the identity of those strains which do not yield clear results by serological methods. The suitability of the probe will depend upon its specificity for target sequences in *L.pneumophila* compared to sequences present in other organisms. To take advantage of the specificity of the probe any background signal generated in the test must be relatively small.

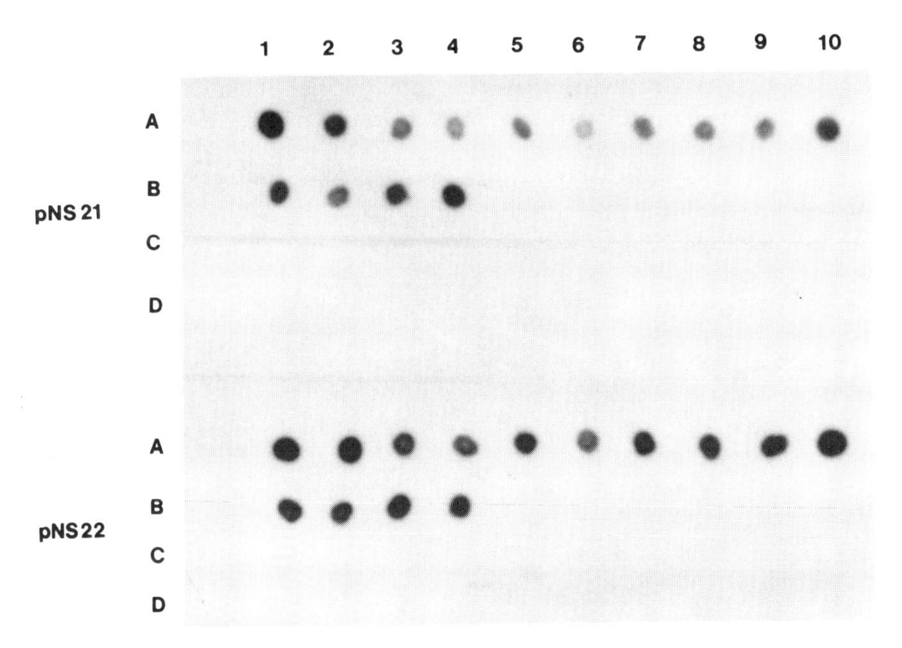

Figure 10.1 Plasmids pNS21 and pNS22, which have inserts of *L.pneumophila* DNA of 3 and 7 kb respectively, were linearized by treatment with restriction endonuclease Pst I and used as templates for T7 RNA polymerase supplied with ^{32}P-labelled UTP substrate. The resulting RNA transcripts were used (at a concentration of 50 ng/ml) to probe a dot blot of purified DNAs (200 ng/dot). The blots were washed under stringent conditions and subjected to autoradiography with intensifying screens at −70 °C. DNA samples were prepared from the following organisms: A1, *L.pneumophila* serogroup 1 (Knoxville-1); A2, serogroup 1 (Philadelphia-1); A3, serogroup 1 (Bellingham-1); A4, serogroup 2 (Togus-1); A5, serogroup 3 (Bloomington-1); A6, serogroup 5 (Dallas-1E); A7, serogroup 5 (Cambridge-2); A8, serogroup 6 (Oxford-1); A9, serogroup 6 (Chicago-2); A10, serogroup 7 (Chicago-8); B1, serogroup 8 (Concord-3); B2, serogroup 9 (IN23-GI-C2); B3, serogroup 10 (Leiden-1); B4, serogroup 11 (797-PA-H); B5, *L.longbeachae* serogroup 1; B6, *L.anisa*; B7, *L.micdadei*; B8, *L.cherrii*; B9, *L.bozemanii*; B10, *L.feeleii*; C1, *L.jordanis*; C2, *L.santicrucis*; C3, *L.dumoffii*; C4, *P.aeruginosa*; C5, *E.coli* K1; C6, *H.influenzae*; C7, *N.meningitidis*; C8, *S.pneumoniae*; C9, *Listeria monocytogenes*; C10, '*Legionella nautarum*' (LC218a); D1, '*L.nautarum*' (LC219a); D2, '*L.nautarum*' (LC220a); D3, '*L.nautarum*' (LC221a); D4, *Klebsiella pneumoniae*; D5, *E.coli*.

As discussed above probes such as the Gen-Probe product, which are derived from *L.pneumophila* rRNA gene sequences, have broad specificity and are capable of forming stable hybrids with the rRNAs or the rRNA genes from all species of *Legionella* as well as with species belonging to unrelated genera.

A difficulty in selecting a specific probe for the Legionellaceae is, therefore, to exclude sequences homologous to the rRNAs of other genera. We have selected a clone from the library by screening with complementary DNA (cDNA) specific for rRNA derived by reverse transcription of purified 16 and 23S rRNAs using M-MLV reverse transcriptase, random primers and a mixture of labelled deoxynucleotides. Several clones were identified as carrying genes for rRNA using plaque filter hybridization (Maniatis *et al.*, 1982) and one λNS142 was selected for further study. This clone was shown to include both 16 and 23S sequences by northern blotting (Saunders *et al.*, 1988). Fragments of the cloned rRNA gene having suitable specificity were selected using the following procedure. The cloned DNA insert was purified, cleaved with a restriction endonuclease (Sau 3A) and the fragments were ligated to the plasmid pT7-1 at its Bam HI site which is located in the polylinker downstream of the bacteriophage T7 promotor. Some of the clones bearing recombinant plasmid were disrupted and the plasmids subjected to agarose gel electrophoresis. The plasmid bands were transferred by capillary blotting to a nylon membrane and probed with cDNAs made by reverse transcription of rRNA from *L.pneumophila*, *E.coli* and *P.aeruginosa*. Plasmids bearing rRNA gene fragments hybridized strongly to the *L.pneumophila* cDNA probe. The degree of hybridization of these plasmids with the cDNA probes from *E.coli* and *P.aeruginosa* varied and it was possible to select several clones for further study on the basis of their especially weak cross-hybridization. RNA transcripts of the rRNA gene insert of one clone pNS9 were prepared using T7 RNA polymerase in the presence of [^{32}P]UTP and a standard mix. Figure 10.2 shows a dot blot of purified DNA from various legionellae and non-legionellae probed with these transcripts.

The probe sequence of pNS9 is short (approximately 300 basepairs) and it is envisaged that the sensitivity of the assay can be increased at least ten-fold by using transcripts derived from plasmid constructs containing multimers of this probe sequence. Such RNA transcripts consisting of repeat sequences will increase the sensitivity of the assay since typically the maximum signal obtainable from a uniformly labelled probe is proportional to its size. It is hoped that the low sequence complexity of the probe will result in rapid hybridization with target sequences. The probe sequence can be cloned in either of the two possible orientations adjacent to the T7 RNA polymerase promotor in this system. The anti-sense strand (complementary to rRNA) will have the advantage of being able to hybridize to rRNA which is present in high copy number in growing cells.

C Probes for typing of strains

The species *L.pneumophila* consists at present of fourteen serogroups which can readily be distinguished, however the majority of cases of Legionnaires' disease are caused by serogroup 1 strains. Consequently a reliable method of typing strains of *L.pneumophila* capable of distinguishing a large number of stable subtypes is required. Such a scheme would allow epidemiological studies to be undertaken enabling cases to be linked and potential sources of infection traced. An ideal scheme would be independent of, and complement, the existing serological markers.

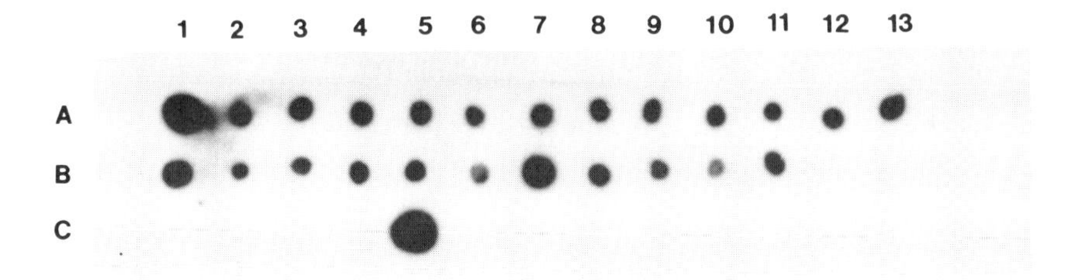

Figure 10.2 Plasmid pNS9, which has a short insert (approximately 300 bp) from the rRNA gene of *L.pneumophila*, was linearized by treatment with restriction endonuclease Pst I and used as a template for T7 RNA polymerase supplied with ^{32}P-labelled UTP substrate. The resulting RNA transcripts were used (at a concentration of 50 ng/ml) to probe a dot blot of purified DNAs (200 ng/dot). The blots were washed under stringent conditions and subjected to autoradiography with intensifying screens at −70 °C. DNA samples were prepared from the following organisms: A1, plasmid T7-1; A2, *L.pneumophila* serogroup 1 (Knoxville-1); A3, serogroup 1 (Philadelphia-1); A4, serogroup 6 (Chicago-2); A5, serogroup 7 (Chicago-8); A6, serogroup 2 (Togus-1); A7, serogroup 6 (Oxford-1); A8, serogroup 1 (Bellingham-1); A9, serogroup 3 (Bloomington-1); A10, serogroup 5 (Cambridge-2); A11, serogroup 5 (Dallas-1E); A12, serogroup 8 (Concord-3); A13, serogroup 9 (IN23-GI-C2); B1, serogroup 10 (Leiden-1); B2, serogroup 11 (797-PA-H); B3, *L.bozemanii* serogroup 1; B4, *L.dumoffii*; B5, *L.santicrucis*; B6, *L.cherrii*; B7, *L.anisa*; B8, *L.longbeachae* serogroup 1; B9, *L.micdadei*; B10, *L.jordanis*; B11, *L.feeleii* serogroup 1; B12, *E.coli*; B13, *P.aeruginosa*; C1, *H.influenzae*; C2, *N.meningitidis*; C3, *S.pneumoniae*; C4, *Listeria monocytogenes*; C5, plasmid T7-1.

The information gained by typing *L.pneumophila* strains may also clarify the taxonomic relationships of the various species. It may also be possible to distinguish virulent from non-virulent strains which would be particularly valuable, for example, in assessing the contamination of water systems.

Probe sequences can be used to detect differences between the DNA sequences of strains by comparing their restriction fragment patterns. Briefly, DNA purified from the test strain is digested with a restriction endonuclease and subjected to agarose gel electrophoresis. The separated DNA fragments are transferred to a nylon membrane by capillary blotting (Southern, 1975) and hybridized to a probe sequence biotin-labelled by primer extension with Klenow fragment (Maniatis *et al.*, 1982). The membrane is washed under non-stringent conditions, to allow some base mismatching of the probe/target duplex. Probe remaining bound to the membrane is then detected using a streptavidin-alkaline phosphatase detection system (Gibco Ltd). The number of bands observed using this procedure is largely a function of the size of the probe sequence and the enzyme used to digest the target DNA. The most useful system developed to date in this laboratory gives variable patterns of about ten clearly defined bands depending on the test strain. Figure 10.3 shows a blot obtained by digestion of DNA from a selection of *L.pneumophila* serogroup 1 isolates. The DNA was treated with the enzyme Nci I which cleaves at the sequence 5′-CC $\binom{G}{C}$ GG-3′. This enzyme was selected because the bands generated are predominantly in the size range 250 – 5000 bp. This is important since if the DNA fragments are too large the number of RFLPs will be small while fragments smaller than 200 bp will not be detected for two reasons: (1) binding to the filter is less efficient; and (2) since the target sequence is short the potential for binding of the probe is small. The probes were chosen by testing a number of cloned DNA fragments. A combination of two unrelated 10 kb sequences are used. An important feature of the probes is that they hybridize with DNA sequences present in all *L.pneumophila* strains examined to date.

This method has considerable advantages over the previously described method for *L.pneumophila* strain identification by examination of the total banding patterns produced by restriction enzyme digestion of DNA from the test organisms (van Ketel *et al.*, 1984). This is because in practice it is very difficult to obtain well-resolved bands due to the complexity of the system in which many hundreds of DNA fragments are present. The demonstration of genomic variation between strains using probes is much more tolerant of minor differences in handling between samples in spite of the greater number of steps involved. Finally, interpretation of the results is easier because a smaller number of bands are examined.

Studies in this and in other laboratories have shown that probes can be successfully used for a variety of taxonomic and epidemiological purposes. It is anticipated that advances in the sensitivity of methods for probe detection will allow an expansion in their use especially in the routine detection of *Legionella* spp. in clinical specimens.

II Monoclonal antibodies

A Introduction

It is a little over ten years since Kohler and Milstein (1975) first described the production of antibodies from a single cell-line using cell-fusion techniques. Since

1975 the technique has been widely applied and the production of monoclonal antibodies (MABs) is now common. At least eight groups have produced MABs against legionellae (Guillet *et al.*, 1983; Joly *et al.*, 1983; McKinney *et al.*, 1983; Para & Plouffe, 1983; Sethi *et al.*, 1983; Watkins & Tobin, 1984; Miyazaki *et al.*, 1985; Harrison, unpublished data). These workers have used MABs in three main areas of study: epidemiology, investigation of virulence markers and the diagnosis of infection.

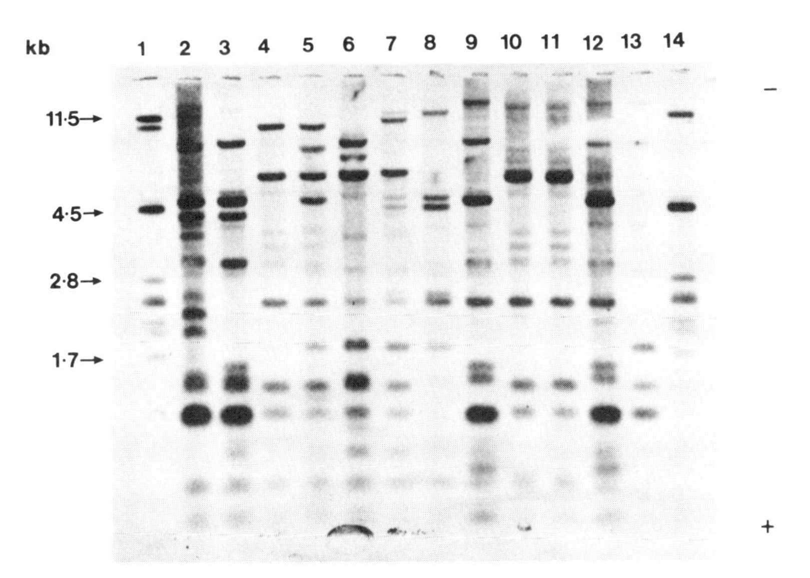

Figure 10.3 DNA isolated from two λ clones, each with inserts of approximately 10 kb of *L.pneumophila* (Knoxville-1) DNA, were labelled with biotin-11-dUTP by primer-extension. The labelled DNAs were mixed in equal proportion of biotinylated residues and used at a concentration of 1 μg/ml, to probe a Southern blot of Nci I digested DNAs from *L.pneumophila* serogroup 1 isolates (1μg/strain) run on an agarose gel (0.8%). The blot was washed under non-stringent conditions. Tracks (1) and (14) show a Pst I digest of λ DNA. The strains examined were run in tracks (2–13) as follows; (2), Knoxville-1; (3), A100/81, environmental, Australia; (4), LC20, clinical, France; (5), LC28, clinical, France; (6), LC36, clinical, France; (7), A32/81, environmental, UK; (8), Pontiac-1; (9), Bellingham-1; (10), LC62, clinical, Holland; (11), LC63, environmental, Holland; (12), Washington-1; (13), A176/79, clinical, UK.

B Epidemiological studies

The reasons for attempting to develop a subgrouping scheme for *L.pneumophila* have been discussed above. At present the most widely reported method is that of immunofluorescence using various panels of MABs. The use of MABs in this way offers several advantages over conventional typing methods, e.g. phage-typing or bacteriocins. These include:

1. The availability of large quantities of reagents of a consistent quality.
2. The rapidity with which results can be obtained.
3. The subgroup can often be determined without isolation of the organism.

This last point is illustrated by several studies. Bibb and colleagues (1984) demonstrated that the subgroup of the infecting organism could be determined by ELISA using urine specimens from patients with LD. Recently the subgroup of *L.pneumophila* serogroup 1 bacilli in post-mortem lung tissue has been determined using either immunoperoxidase staining techniques (Theaker *et al.*, 1987) or, in this laboratory, by immunofluorescence (authors' unpublished data).

Although MAB subgrouping has been attempted for all serogroups of *L.pneumophila* most of the published data concerns serogroup 1. McKinney and colleagues (1983) selected 9 clones, out of approximately 1000 hybridomas, which had a range of different specificities. The authors examined 130 isolates of *L.pneumophila* serogroup 1 and found that they could be divided into two distinct subgroups comprising 13 different staining patterns. Several sets of isolates which were thought to be epidemiologically associated (e.g. multiple isolates from a single outbreak) were found to give the same staining pattern. This study indicated the scale of antigenic diversity within *L.pneumophila* serogroup 1 and demonstrated the potential of such a system for identifying a common source of infection in outbreaks of LD.

Later in the same year Joly and colleagues (1983) published the results obtained using another panel of MABs. These workers identified seven antigen patterns using a panel of eight MABs. Although in this study only a small number of isolates were examined the results again demonstrated the potential of this technique. In a third study, by Watkins and Tobin (1984), ten MABs were used to classify isolates into one of seven subgroups. Later these workers refined their set of MABs to give three major subgroups, two of which were further subdivided into minor groups (Watkins *et al.*, 1985).

In each of the above studies the authors demonstrated that epidemiologically related isolates were indistinguishable by MAB subgrouping. However, the significance of such results is as yet uncertain. Until these subgrouping schemes are thoroughly evaluated it cannot be assumed that two isolates having different staining patterns (which are therefore distinguishable) are not representatives of the same strain. This has been highlighted by data from several studies. Joly and colleagues found that a single strain of *L.pneumophila* could give different staining patterns using the same MAB. They found that in some cases this was dependent on the passage history of the isolate while in others no explanation was discovered (Joly & Ramsay, 1985). Similarly Watkins and colleagues (1985) noted that variation in the methods used to prepare the strains for examination could significantly modify the staining pattern obtained.

In addition to the technical limitations on reproducibility the stability of the antigens against which the MABs are directed also requires evaluation. Watkins and colleagues (1985) reported that a strain subcultured repeatedly over 4–6 weeks retained the same staining pattern suggesting that the antigens are stable. However, more recently it has been demonstrated that the multiple subculture of a strain of *L.pneumophila* serogroup 1 caused a quantitative (but not qualitative) decrease in the number of surface located high-affinity epitopes against which one MAB was directed (Petitjean *et al.*, 1986). Since the loss or gain of a particular epitope could entirely alter the subgroup of an isolate, MABs used for subgrouping must be directed against antigens which are stable over a wide range of environmental conditions. This aspect of subgrouping using MABs has not yet been sufficiently investigated.

The most promising MAB subgrouping scheme for *L.pneumophila* serogroup 1 so far developed was produced as a collaborative project by the three groups of workers whose work has already been discussed (Joly *et al.*, 1986). The seven most useful MABs from the three previous panels were incorporated into a standard panel and a defined protocol was developed. The authors selected these particular MABs because they gave ten clearly defined subgroups (Table 10.2) and showed no major variations in staining intensity with different isolates.

Table 10.2 Subgroups of *L.pneumophila* identified with the seven monoclonal antibodies selected from the different laboratories. (Reproduced with permission of the American Society of Microbiology from Joly *et al.*, 1986b.)

Representative strain	Indirect immunofluorescence assay staining intensity with:[a]						
	MAB1	MAB2	MAB3	W32	33G12	32A12	144C2
Philadelphia-1	3	3	0	0	3	3	0
Allentown-1	3	3	0	0	3	0	0
Benidorm 030E	3	3	0	0	3	0	3
Knoxville-1	3	3	3	0	0	V[b]	0
France 5811	3	3	0	0	0	0	0
OLDA	3	0	0	0	0	3	3
Oxford 4032E	3	0	0	0	0	3	0
Heysham-1	3	0	3	0	0	3	3
Camperdown-1	3	0	0	0	0	0	0
Bellingham-1	3	0	0	3	0	0	3

a. Staining intensity: 3. bright fluorescence; 2. good fluorescence; 1. barely visible fluorescence; 0. no fluorescence.

b. Variable results were obtained with this monoclonal antibody and the different *L.pneumophila* serogroup 1 strains belonging to this subgroup.

In this laboratory we have subgrouped over 200 strains using this international panel of MABs and have encountered no major problems. With one exception all strains have been identified as belonging to one of the ten subgroups so far described. It is noteworthy that in several instances an isolate which at first failed to fit into the existing scheme was subsequently found to be a mixture of two isolates of different subgroups derived from a single source.

It is hoped that with all three laboratories now using the same panel of MABs, this subgrouping scheme will soon be thoroughly evaluated allowing its great potential as a powerful epidemiological tool to be realized.

In addition to MABs directed against somatic antigens we have produced a number of anti-flagella MABs. Depending on their reactivity with three of these MABs, flagellate isolates of *L.pneumophila* can be subtyped into one of four groups. These subtypes are apparently independent of the somatic subgroups so it should be possible to subdivide *L.pneumophila* serogroup 1 isolates still further. Furthermore, since the flagellar subtype is also independent of serogroup this could be used to provide a limited number of subtypes for isolates of serogroups other than serogroup 1. Examples of the flagellar MAB patterns encountered to date with isolates of each subgroup and serogroup are shown in Table 10.3. These studies are at a preliminary stage and factors such as subtype stability, reproducibility, discrimination and possible phase variation have not yet been investigated. Thus it is too early to predict whether flagellar subtyping will be of any practical value.

C Virulence markers

Although *L.pneumophila* serogroup 1 strains are ubiquitous and frequently isolated from man-made habitats, LD is a relatively uncommon illness, even among those groups of individuals most at risk. One possible explanation for this is that different strains vary in their virulence.

The first evidence supporting this theory was reported by Brown and colleagues (1982) who made a detailed examination of isolates associated with a nosocomial outbreak of LD. Two populations of bacteria were recognized depending upon the carriage or absence of a 80 megaDalton plasmid. The plasmid-containing bacteria were present in the hospital water before and after the outbreak of LD, however the plasmidless organisms were only detectable before and during the outbreak. Moreover, all 23 clinical isolates obtained during the outbreak were plasmidless, and wards from which plasmidless organisms had been recovered had a significantly higher number of LD cases than did other wards.

The association between an isolate's virulence and its reaction with a particular MAB was first noted by Plouffe and colleagues (1983). Twenty cases of LD caused by *L.pneumophila* serogroup 1 were diagnosed in a hospital in Columbus, USA. The cases were associated with two buildings in the hospital, UH and RH.

Table 10.3 Subtypes of *L.pneumophila* identified with three anti-flagella monoclonal antibodies.

Pattern no.	Flagella MAB FL7	FL3	FL8	Strains from each *L.pneumophila* subgroup with this pattern[1]	Strains of each *L.pneumophila* serogroup with this pattern
1	+	+	+	All subgroups	Serogroups 1 – 10
2	+	+	–	Philadelphia, Knoxville, Oxford	Serogroup 1
3	+	–	+	Philadelphia, Allentown, Benidorm, Knoxville, Bellingham	Serogroups 1, 3, 4, and 8
4	+	–	–	Benidorm	Serogroups 1, 2, 7, and 8

1. Subgroups as defined in Table 10.2.

Although the majority of susceptible patients were housed in RH, only one case of LD was detected here as compared with 19 cases in UH. The buildings were supplied with the same water but their plumbing systems were separate. Isolates, which were obtained with a similar frequency from both buildings, could be separated into three subgroups dependent upon their reaction with a MAB, their colony morphology, and their plasmid content. Only subgroup RH-1 isolates were found in building RH and this subgroup was also isolated from the single RH patient. Of the isolates in building UH 99% were subtype UH-1 or UH-2 as were all 15 clinical isolates from UH patients.

The association seen in this study between the virulence of an isolate and its reactions with a MAB could have been coincidental, however recent data (Dournon *et al.*, 1986a) provide an alternative explanation.

Dournon and colleagues (1986a) examined a series of nearly 200 apparently unrelated clinical and environmental isolates using the MAB subgrouping scheme described by McKinney and colleagues (1983). Most clinical isolates (95%) reacted with MAB2 (MAB2+) compared to only 35% of environmental isolates. Only five patients were infected by strains which did not react with MAB2 (MAB2−) and four of these were end-stage immunocompromised patients receiving 'aggressive' chemotherapy. This contrasts with the observation that more than half the patients infected with MAB2+ strains were apparently healthy. From these studies it can be concluded that MAB2+ strains are more virulent and likely to cause LD than are MAB2− strains. These data are supported by observations at the CDC and in this laboratory, that all outbreak-associated *L.pneumophila* serogroup 1 isolates are MAB2+ strains.

Further studies are needed to confirm and extend this work which indicates that it may be possible to identify environmental sites containing *L.pneumophila* strains likely to cause disease from the more frequently encountered sites harbouring relatively non-pathogenic strains.

D Diagnosis of infection

The high specificity and potentially unlimited supply of monoclonal antibodies for use as diagnostic tools can be expected to lead to significant improvements in the laboratory diagnosis of *Legionella* infections. To date very few studies using MABs for this purpose have been reported. Guillet and colleagues (1983) demonstrated that antigen could be detected in bronchial secretions from a culture-proven LD patient. Similarly Miyazaki and colleagues (1985) used an immunoperoxidase technique to reveal legionellae in human lung tissue. As both studies involved only single patients nothing can be concluded concerning the utility of such techniques.

The factor limiting the suitability of most MABs for use in diagnostic assays is their restricted range of reactivity. Several workers have been able to detect all isolates of *L. pneumophila* serogroup 1 by preparing pools of two or three MABs (McKinney *et al.*, 1983; Watkins & Tobin, 1984). However, to extend this reactivity to all serogroups of *L. pneumophila* a substantial number of MABs have to be pooled (Joly *et al.*, 1983), increasing the complexity of preparation and the need for careful quality control of the reagents.

In 1984 Gosting and colleagues reported the production of a MAB directed against a species-specific antigen. This antigen is a heat-stable major outer membrane protein with a molecular weight of about 25 000 – 29 000 Daltons and is present in strains of all serogroups of *L. pneumophila*. The antigen is only partly surface-exposed and consequently an immunofluorescent assay using this MAB resulted in only partial staining of the cells. Pretreatment of the bacterial cells with a detergent/EDTA solution overcame this limitation and the MAB is now commercially available (Genetic Systems Corp.) in kit form as a FITC-conjugated reagent. The conjugate preparation has been reformulated so that addition of detergent to the sample is no longer required. This kit is extremely useful as a reagent for the confirmation of isolates as *L. pneumophila*. Two studies using the kit found that it correctly identified all *L. pneumophila* isolates examined and gave no false-positive reactions with organisms reported to react with equivalent polyvalent reagents (Tenover *et al.*, 1985; Fallon, 1986).

The clinical utility of the kit has also been evaluated (Edelstein *et al.*, 1985; see Chapter 8). These authors examined 24 lower respiratory tract specimens from culture-proven cases of LD. All 24 specimens were positive using both the Genetic Systems MAB and CDC polyvalent rabbit antisera. The authors concluded that the theoretically superior specificity of a MAB and the ease of its use made this the reagent of choice. Unfortunately the utility of the MAB is limited as it cannot be used successfully on formalin-fixed or histopathologically prepared specimens. Although this MAB can only be obtained from one manufacturer at present, other manufacturers will undoubtedly produce similar reagents in the future.

A Laboratory Manual for *Legionella*
Edited by T. G. Harrison and A. G. Taylor

APPENDIX 1

Legionella Strains Available from the NCTC and ATCC

Species/Serogroup		Collection[1] NCTC	ATCC	Original designation	References
Legionella pneumophila	Sgp1	11192	33152	Philadelphia-1[2]	Brenner *et al.* (1979)
		11286	33153	Knoxville-1	
		12024	43016	Allentown-1	
		12006	43108	Benidorm 030E	
		12007	43112	France 5811	
		12008	43109	OLDA	Joly *et al.* (1986b)[3]
		12009	43110	Oxford 4032E	
		12025	43107	Heysham-1	
		12098	43113	Camperdown-1	
		11404	43111	Bellingham-1	
		11191		Pontiac-1	
	Sgp2	11230	33154	Togus-1	McKinney *et al.* (1979b)
	Sgp3	11232	33155	Bloomington-2	Morris *et al.* (1979)
	Sgp4	11233	33156	Los-Angeles-1	McKinney *et al.* (1979a)
	Sgp5	11417		Cambridge-2	Nagington *et al.* (1979a)
		11405	33216	Dallas-1E	England *et al.* (1980)
	Sgp6	11287		Oxford-1	Tobin *et al.* (1980)
		11406	33215	Chicago-2	McKinney *et al.* (1980)
	Sgp7	11984	33823	Chicago-8	Bibb *et al.* (1983)
	Sgp8	11985	35096	Concord-3	Bisset *et al.* (1983)
	Sgp9	11986	35289	IN-23-G1-C2	Edelstein *et al.* (1984a)
	Sgp10	12000	43283	Leiden-1	Meenhorst *et al.* (1985)
	Sgp11	12179	43130	797-PA-H	Thacker *et al.* (1986)
	Sgp12	12180	43290	570-CO-H	Thacker *et al.* (1987)
	Sgp13	12181	43736		
	Sgp14	12174		1586-SCT-H	Benson *et al.* (1988)
			43703	1169-MN-H	

Species/Serogroup		Collection[1] NCTC	ATCC	Original designation	References
Legionella anisa		11974	35292	WA-316-C3	Gorman *et al.* (1985)
Legionella birminghamensis			43702	1407-AL-H	Wilkinson *et al.* (1987)
Legionella bozemanii	Sgp1	11368	33217	WIGA	Brenner *et al.* (1980)
	Sgp2	11975	35545	Toronto-3	Tang *et al.* (1984)
Legionella cincinnatiensis			43753	72-OH-H	
Legionella cherrii		11976	35252	ORW	Brenner *et al.* (1985)
Legionella dumoffii		11370	33279	NY-23	Brenner *et al.* (1980)
Legionella erythra		11977	35303	SE-32A-C8	Brenner *et al.* (1985)
Legionella feeleii	Sgp1	12022	35072	WO-44C	Herwaldt *et al.* (1984)
	Sgp2	11978	35849	691-WI-H	Thacker *et al.* (1985b)
'*Legionella geestiae*'					Dennis *et al.*[4]
Legionella gormanii		11401	33297	LS-13	Morris *et al.* (1980)
Legionella hackeliae	Sgp1	11979	35250	Lansing-2	Brenner *et al.* (1985)
	Sgp2	11980	35999	798-PA-H	Wilkinson *et al.* (1985)
Legionella israelensis		12010	43119	Bercovier-4	Bercovier *et al.* (1986)
Legionella jamestowniensis		11981	35298	JA-26-G1-E2	Brenner *et al.* (1985)
Legionella jordanis		11533	33623	BL-540	Cherry *et al.* (1982)
'*Legionella londoniensis*'					Dennis *et al.*[4]
Legionella longbeachae	Sgp1	11477	33462	Long Beach-4	McKinney *et al.* (1981)
	Sgp2	11530	33484	Tucker-1	Bibb *et al.* (1981)
Legionella maceachernii		11982	35300	PX-1-G2-E2	Brenner *et al.* (1985)
Legionella micdadei		11371	33218	TATLOCK	Hébert *et al.* (1980)
'*Legionella nautarum*'					Dennis *et al.*[4]
Legionella oakridgensis		11531	33761	OR-10	Orrison *et al.* (1983b)
Legionella parisiensis		11983	35299	PF-209C-C2	Brenner *et al.* (1985)
'*Legionella quateriensis*'					Dennis *et al.*[4]
Legionella rubrilucens		11987	35304	WA-270A-C2	Brenner *et al.* (1985)
Legionella sainthelensi		11988	35248	Mt St Helens-4	Campbell *et al.* (1984)
Legionella santicrucis		11989	35301	SC-63-C7	Brenner *et al.* (1985)
Legionella spiritensis		11990	35249	Mt St Helens-9	Brenner *et al.* (1985)
Legionella steigerwaltii		11991	35302	SC-18-C9	Brenner *et al.* (1985)
Legionella wadsworthii		11532	33877	81-716A	Edelstein *et al.* (1982a)
'*Legionella worsliensis*'					Dennis *et al.*[4]

1. Catalogue numbers of strains available from either the National Collection of Type Cultures (NCTC) or American Type Culture Collection (ATCC).
2. Philadelphia-1 is the type strain.
3. The bracketed strains are representative strains of the MAB subgroup scheme described by Joly and colleagues (1986b).
4. Proposed new species, Dennis *et al.* manuscript in preparation, type strains to be deposited with the NCTC and ATCC shortly.

A Laboratory Manual for *Legionella*
Edited by T. G. Harrison and A. G. Taylor

APPENDIX 2

Preservation of *Legionella* on Glass Beads

Although freeze-drying in glass ampoules is the preferred method for long-term storage of bacteria it is time consuming and laborious for the routine storage of laboratory isolates. However, it is important that strains of legionellae be reliably stored after a minimal number of passages on laboratory media. The storage of legionellae on glass beads has proved very successful in this laboratory and the method (Feltham *et al.*, 1978) is briefly outlined below. A detailed description of this and other suitable storage methods can be found in *Maintenance of Microorganisms: A Manual of Laboratory Methods* (Kirsop & Snell, 1984).

Preparation of beads
Embroidery beads (2 mm, Ellis & Farrier Ltd) are soaked in 5% HCl overnight and then washed repeatedly in tap water until the pH is that of tap water. After a final rinse in distilled water the beads are dried overnight in a drying cabinet. Beads (20 – 30) are placed in a suitable screw-cap container, loosely capped, and sterilized by autoclaving. The lids are then tightened for storage.

Preparation of glycerol broth

Oxoid nutrient broth No. 2	5 g
Glycerol	30 ml
Distilled water	170 ml

The nuturient broth powder is dissolved in distilled water and the glycerol added. The broth is dispensed into 2 ml volumes sterilized by autoclaving, and stored at room temperature until required.

Procedure

Growth from a BCYE plate, incubated for 48–72 hours, is emulsified in about 1 ml of glycerol broth to give a dense bacterial suspension. The suspension is added to a vial which is then gently shaken to ensure that the beads are thoroughly coated. Excess fluid is then withdrawn from the base of the vial with a Pasteur pipette. The beads are sloped in the vials to facilitate their removal after freezing. Cultures can then be stored either in liquid nitrogen or at −70 °C until required.

A bead should be recovered immediately after freezing to check for purity and viability of the stored bacterial strain.

Recovery of strains from beads

To avoid the repeated thawing and freezing of the stored culture it is essential that the vial containing the beads remains frozen throughout the recovery procedure. To facilitate this the vials can be placed in a container at −70 °C (e.g. a frozen paraffin block) while they are removed from the liquid nitrogen. Sterile forceps are used to remove a single bead from the vial. The bead is then placed on to a BCYE plate and a flamed wire-loop is used to roll it over the surface of the agar. The vial containing the remaining beads is immediately replaced in the liquid nitrogen. The inoculated BCYE plate is incubated for 48–72 hours at 37 °C, at which time bacterial growth should be evident.

It has been reported that the beads supplied by some manufacturers may be unsuitable for the storage of legionellae (Barker & Till, 1986). The beads were being used to inoculate stock strains of legionellae into a liquid medium but it was found that growth was inhibited. The leaching of sodium ions from the glass beads into the medium was suggested as a possible explanation for the failure of the organisms to grow. Using the method described above where the bacteria are recovered on to solid media we have not encountered any similar problems.

A Laboratory Manual for *Legionella*
Edited by T. G. Harrison and A. G. Taylor

Bibliography

Ampel, N. M, Ruben, F. L, & Norden, C. W. (1985). Cutaneous abscess caused by *Legionella micdadei* in an immunosuppressed patient. *Ann. Intern. Med.*, **102**:630–2.
Arnow, P. M., & Gardner, P. (1983). Usefulness of the Gram stain in Legionnaires' disease. *Microbiol. Newsl.*, **5**:125–7.
Arnow, P. M, Chou, T., Weil, D., Shapiro, E. N., & Kretzschmar, C. (1982). Nosocomial Legionnaires' disease caused by aerosolized tap water from respiratory devices. *J. Infect. Dis.*, **146**:460–7.
Arnow, P. M., Boyko, E. J., & Friedman, E. L. (1983). Perirectal abscess caused by *Legionella pneumophila* and mixed anaerobic bacteria. *Ann. Intern. Med.*, **98**:184–5.
Ashworth, J., & Colbourne, J. S. (1987). The testing of non-metallic materials for use in contact with potable water, and the inter-relationships with in service use. In: Hopton, J. W., & Hill, E. C. (eds), *Industrial Microbiological Testing*. Oxford: Blackwell Scientific (SAB Technical Series No. 23), pp.151–70.
Baine, W. B., Rasheed, J. K., Feeley, J. C., Gorman, G. W., & Casida, L. E. (1978). Effect of supplemental L-tyrosine on pigment production in cultures of the Legionnaires' disease bacterium. *Curr. Microbiol.*, **1**:93–4.
Baptiste-Desruisseaux, D., Duperval, R., & Marcoux, J. A. (1985). Legionnaires' disease in the immunocompromised host: usefulness of Gram's stain. *Can. Med. Assoc. J.*, **133**:117–18.
Barka, N., Tomasi, J. P., & Stadtsbaeder, S. (1986). Elisa using whole *Legionella pneumophila* cell as antigen: comparison between monovalent and polyvalent antigens for the serodiagnosis of human legionellosis. *J. Immunol. Methods*, **93**:77–81.
Barker, J., & Till, D. H. (1986). Survival of *Legionella pneumophila*. *Med. Lab. Sci.*, **43**:388–9.
Bartlett, C. L. R., Kurtz, J. B., Hutchison, J. G. P., Turner, G. C., & Wright, A. E. (1983). Legionella in hospital and hotel water supplies. *Lancet*, **ii**:1315.
Benson, R. F., Thacker, W. L., Wilkinson, H. W., Fallon, R. J., & Brenner, D. J. (1988). *Legionella pneumophila* serogroup 14 isolated from patients with fatal pneumonia. *J. Clin. Microbiol.* **26**:382.
Bercovier, H., Steigerwalt, A. G., Derhi-Cochin, M., *et al.* (1986). Isolation of legionellae from oxidation ponds and fishponds in Israel and description of *Legionella israelensis* sp. nov. *Int. J. Sys. Bacteriol.*, **36**:368–71.

Berdal, B. P., Farshy, C. E., & Feeley, J. C. (1979). Detection of *Legionella pneumophila* antigen in urine by enzyme-linked immunospecific assay. *J. Clin. Microbiol.*, **9**:575–8.

Bibb, W. F., Sorg, R. J., Thomason, B. M., *et al.* (1981). Recognition of a second serogroup of *Legionella longbeachae*. *J. Clin. Microbiol.*, **14**:674–7.

Bibb, W. F., Arnow, P. M., Dellinger, D. L., & Perryman, S. R. (1983). Isolation and characterization of a seventh serogroup of *Legionella pneumophila*. *J. Clin. Microbiol.*, **17**:346–8.

Bibb, W. F., Arnow, P. M., Thacker, L., & McKinney, R. M. (1984). Detection of soluble *Legionella pneumophila* antigens in serum and urine specimens by enzyme-linked immunosorbent assay with monoclonal and polyclonal antibodies. *J. Clin. Microbiol.*, **20**:478–82.

Bissett, M. L., Lee, J. O., & Lindquist, D. S. (1983). New serogroup of *Legionella pneumophila*, serogroup 8. *J. Clin. Microbiol.*, **17**:887–91.

Bopp, C. A., Sumner, J. W., Morris, G. K., & Wells, J. G. (1981). Isolation of *Legionella* spp. from environmental water samples by low-pH treatment and use of a selective medium. *J. Clin. Microbiol.*, **13**:714–19.

Brabender, W., Hinthorn, D. R., Asher, M., Lindsey, N. J., & Liu, C. (1983). *Legionella pneumophila* wound infection. *JAMA*, **250**:3091–2.

Bradstreet, C. M. P., & Taylor, C. E. D. (1962). Technique of complement-fixation test applicable to the diagnosis of virus disease. *Monthly Bulletin of the Ministry of Health and the Public Health Laboratory Service*, **21**:96–104.

Brenner, D. J. (1986). Classification of Legionellaceae: Current status and remaining questions. *Isr. J. Med. Sci.*, **22**:620–32.

Brenner, D. J., Steigerwalt, A. G., & McDade, J. E. (1979). Classification of the Legionnaires' disease bacterium: *Legionella pneumophila*, genus novum, species nova, of the family *Legionellaceae* familia nova. *Ann. Intern. Med.*, **90**:656–8.

Brenner, D. J., Steigerwalt, A. G., Gorman, G. W., *et al.* (1980). *Legionella bozemanii* sp. nov. and *Legionella dumoffii* sp. nov.: classification of two additional species of *Legionella* associated with human pneumonia. *Curr. Microbiol.*, **4**:111–16.

Brenner, D. J., Feeley, J. C., & Weaver, R. E. (1984). Family VII *Legionellaceae*. In: Kreig, N. R., & Holt, J. G. (eds), *Bergey's Manual of Systematic Bacteriology*, vol. 1. Baltimore: Williams and Wilkins, pp.279–88.

Brenner, D. J., Steigerwalt, A. G., Gorman, G. W., *et al.* (1985). Ten new species of *Legionella*. *Int. J. Sys. Bacteriol.*, **35**:50–9.

Broome, C. V., & Fraser, D. W. (1979). Epidemiologic aspects of legionellosis. *Epidemiol. Rev.*, **1**:1–15.

Brown, A., Vickers, R. M., Elder, E. M., Lema, M., & Garrity, G. M. (1982). Plasmid and surface antigen markers of endemic and epidemic *Legionella pneumophila* strains. *J. Clin. Microbiol.*, **16**:230–5.

Brown, S. L., Bibb, W. F., & McKinney, R. M. (1984). Restrospective examination of lung tissue specimens for the presence of *Legionella* organisms: Comparison of an indirect fluorescent-antibody system with direct fluorescent-antibody testing. *J. Clin. Microbiol.*, **19**:468–72.

Brown, A., Lema, M. W., & Brown-Schlumpf, M. S. (1986). Antigenic specificity of the antibody response in humans during legionellosis. *Infection*, **14**:108–14.

Buesching, W. J., Brust, R. A., & Ayers, L. W. (1983). Enhanced primary isolation of *Legionella pneumophila* from clinical specimens by low-pH treatment. *J. Clin. Microbiol.*, **17**:1153–5.

Campbell, J., Bibb, W. F., Lambert, M. A., *et al.* (1984). *Legionella sainthelensi* : a new species of *Legionella* isolated from water near Mt St Helens. *Appl. Environ. Microbiol.*, **47**:369–73.

Chandler, F. W., Hicklin, M. D., & Blackmon, J. A. (1977). Demonstration of the agent of Legionnaires' disease in tissue. *N. Engl. J. Med.*, **297**:1218–20.

Cherry, W. B., Pittman, B., Harris, P. P., *et al.* (1978). Detection of Legionnaires' disease bacteria by direct immunofluorescent staining. *J. Clin. Microbiol.*, **8**:329–38.

Cherry, W. B., Gorman, G. W., Orrison L. H., *et al.* (1982). *Legionella jordanis*: a new species of *Legionella* isolated from water and sewage. *J. Clin. Microbiol.*, **15**:290–7.

Chester, B., Poulos, E. G., Demaray, M. J., Albin, E., & Prilucik, T. (1983). Isolation of *Legionella pneumophila* serogroup 1 from blood with nonsupplemented blood culture bottles. *J. Clin. Microbiol.*, **17**:195–7.

Collins, M. D. (1985). Isoprenoid quinone analyses in bacterial classification and identification. In: Goodfellow, M., & Minnikin, D. E. (eds) *Chemical Methods in Bacterial Systematics*. London: Academic Press, pp.267–87.

Collins, M. D., & Gilbart, J. (1983). New members of the Coenzyme Q series from Legionellaceae. *FEMS Microbiol. Letts.*, **16**:251–5.

Collins, M. D., & Jones, D. (1981a). Distribution of isoprenoid quinone structural types in bacteria and their taxonomic implications. *Microbiol. Rev.*, **45**:316–54.

Collins, M. D., & Jones, D. (1981b). A note on the separation of natural mixtures of bacterial ubiquinones using reverse-phase partition thin-layer chromatography and high performance liquid chromatography. *J. Appl. Bacteriol.*, **51**:129–34.

Collins, M. D., Pirouz, T., Goodfellow, M., & Minnikin, D. E., (1977). Distribution of menaquinones in actinomycetes and corynebacteria. *J. Gen. Microbiol.*, **100**:221–30.

Collins, M. T., Cho, S. N., & Reif, J. S. (1982). Prevalence of antibodies of *Legionella pneumophila* in animal populations. *J. Clin. Microbiol.*, **15**:130–6.

Collins, M. T., Espersen, F., Hoiby, N., Cho, S. N., Friis-Moller, A., & Reif, J. S. (1983). Cross-reactions between *Legionella pneumophila* (serogroup 1) and twenty-eight other bacterial species, including other members of the family *Legionellaceae. Infect. Immun.*, **39**:1441–56.

Collins, M. T., McDonald, J., Hoiby, N., & Aalund, O. (1984). Agglutinating antibody titers to members of the family *Legionellaceae* in cystic fibrosis patients as a result of cross-reacting antibodies to *Pseudomonas aeruginosa*. *J. Clin. Microbiol.*, **19**:757–62.

Cutz, E., Thorner, P. S., Rao, C. P., Toma, S., Gold, R., & Gelfand, E. W. (1982). Disseminated *Legionella pneumophila* infection in an infant with severe combined immunodeficiency. *J. Pediatr.*, **100**:760–2.

Department of Health and Social Security *et al.* (1975). Report of the Working Party on the Laboratory Use of Dangerous Pathogens. (Chairman: Sir George Godber.) London: HMSO.

Department of Health and Social Security *et al.* (1978). Code of Practice for the Prevention of Infection in Clinical Laboratories and Post-mortem Rooms. London: HMSO.

Dennis, P. J., Taylor, J. A., & Barrow, G. I. (1981). Phosphate buffered, low sodium chloride blood agar medium for *Legionella pneumophila*. *Lancet* **ii**:636.

Dennis, P. J., Bartlett, C. L. R., & Wright, A. E. (1984a). Comparison of isolation methods for *Legionella* spp. In: Thornsberry, C., Balows, A., Feeley, J. C., & Jakubowski, W. (eds), *Legionella: Proceedings of the 2nd International Symposium*. Washington DC: Am. Soc. Microbiol., pp.294–6.

Dennis, P. J., Green, D., & Jones, B. P. C. (1984b). A note on the temperature tolerance of *Legionella*. *J. Appl. Bacteriol.*, **56**:349–50.

Dennis, P. J., *et al.* Manuscript in preparation.

Desplaces, N., Nahapetian, K., & Dournon, E. (1984). Inventaire des *Legionella* dans l'environnement parisien: implications pratiques. *Le Presse Medicale*, **13**:1875–9.

Dieterle, R. R. (1927). Method for demonstration of *Spirocheata pallida* in single microscopic sections. *Arch. Neurol. Psychiatry*, **18**:73–80.

Dorman, S. A., Hardin, N. J., & Winn, W. C. (1980). Pyelonephritis associated with *Legionella pneumophila*, serogroup 4. *Ann. Intern. Med.*, **93**:835–7.

Doshi, N., & Taylor, A. G. (1984). Comparison of the Vi indirect fluorescent antibody test with the Widal agglutination method in the serodiagnosis of typhoid fever. *J. Clin. Pathol.*, **37**:805–8.

Dournon, E., Buré, A., Kemeny, J. L., Pourriat, J. L., & Valeyre, D. (1982). *Legionella pneumophila* peritonitis. *Lancet*, **i**:1363.

Dournon, E., Mayand, C., Buré, A., Desplaces, N., & Christol, D. (1983). Epidemiological features of Legionnaires' disease in the Paris area. *Zentralbl. Bakteriol. Mikrobiol. Hyg. (A)*, **255**:76–83.

Dournon, E., Desplaces, N., Rajagopalan, P., Bibb, W. F., & McKinney, R.M. (1986a). Recognition of virulent *L.pneumophila* serogroup 1 strains with monoclonal antibodies. *Isr. J. Med. Sci.*, **22**:756.

Dournon, E., Rajagopalan, P., & Assous, M. (1986b). Bacteremia in Legionnaires' disease. *Isr. J. Med. Sci.*, **22**:759.

Dournon, E., Rajagopalan, P., Vildé, J. L., & Pocidalo, J. J. (1986c). Efficacy of pefloxacin in comparison with erythromycin in the treatment of experimental guinea pig legionellosis. *J. Antimicrob. Chemother.*, **17**, Suppl. B:41–8.

Dowling, J. N., McDevitt, D. A., & Pasculle, A. W., (1984a). Disk diffusion antimicrobial susceptibility testing of members of the family *Legionellaceae* including erythromycin-resistant variants of *Legionella micdadei*. *J. Clin. Microbiol.*, **19**:723–9.

Dowling, J. N., Pasculle, A. W., Frola, F. N., Zaphyr, M. K., & Yee, R. B. (1984b). Infections caused by *Legionella micdadei* and *Legionella pneumophila* among renal transplant recipients. *J. Infect. Dis.*, **149**:703–13.

Drucker, D. P., (1981). *Microbiological Applications of Gas Chromatography*. Cambridge: Cambridge University Press.

Dumoff, M. (1979). Direct *in vitro* isolation of the Legionnaires' disease bacterium in two fatal cases: cultural and staining characteristics. *Ann. Intern. Med.*, **90**:694–6.

Durham, T. M., Hale, C. T., Harrell, W. K., & Colvin, H. M. (1984). Commercial *Legionella* direct and indirect immunofluorescence reagents. In: Thornsberry, C., Balows, A., Feeley, J. C., Jakubowski, W. (eds), *Legionella: Proceedings of the 2nd International Symposium*. Washington DC: Am. Soc. Microbiol., pp.27–8.

Edelstein, P. H. (1981). Improved semi-selective medium for isolation of *Legionella pneumophila* from contaminated clinical and environmental specimens. *J. Clin. Microbiol.*, **14**:298–303.

Edelstein, P. H. (1982). Comparative study of selective media for isolation of *Legionella pneumophila* from potable water. *J. Clin. Microbiol.*, **16**:697–9.

Edelstein, P. H. (1983). Culture diagnosis of *Legionella* infections. *Zentralbl. Bakteriol. Mikrobiol. Hyg. (A)*, **255**:96–101.

Edelstein, P. H. (1984a). Laboratory diagnosis of Legionnaires' disease. In: Thornsberry, C., Balows, A., Feeley, J. C. & Jakubowski, W. (eds) *Legionella: Proceedings of the 2nd International Symposium*. Washington DC: Am. Soc. Microbiol., pp.3–5.

Edelstein, P. H. (1984b). *Legionnaires' Disease Laboratory Manual*. Springfield, Va: National Technical Information Service (Document no. PB84-156827).

Edelstein, P. H. (1986). Evaluation of the Gen-Probe DNA probe for the detection of legionellae in culture. *J. Clin. Microbiol.*, **23**:481–4.

Edelstein, P. H., & Finegold, S. M. (1979). Use of a semi-selective medium to culture *Legionella pneumophila* from contaminated lung specimens. *J. Clin. Microbiol.*, **10**:141–3.

Edelstein, P. H., & Meyer, R. D. (1980). Susceptibility of *Legionella pneumophila* to twenty antimicrobial agents. *Antimicrob. Agents. Chemother.*, **18**:403–8.

Edelstein, P. H., & Meyer, R. D. (1984). Legionnaires' disease: a review. *Chest* **85**:114–20.
Edelstein, P. H., & Pryor, E. P. (1985). A new biotype of *Legionella dumoffii*. *J. Clin. Microbiol.*, **21**:641–2.
Edelstein, P. H., Meyer, R. D., & Finegold, S. M. (1979). Isolation of *Legionella pneumophila* from blood. *Lancet* **i**:750–1.
Edelstein, P. H., McKinney, R. M., Meyer, R. D., Edelstein, M. A. C., Krause, C. J., & Finegold, S. M. (1980). Immunologic diagnosis of Legionnaires' disease: cross reactions with anaerobic and microaerophilic organisms and infections caused by them. *J. Infect. Dis.*, **141**:652–5.
Edelstein, P. H., Brenner, D. J., Moss, C. W., Steigerwalt, A. G., Francis, E. M., & George, W. L. (1982a). *Legionella wadsworthii* species nova: a cause of human pneumonia. *Ann. Intern. Med.*, **97**:809–13.
Edelstein, P. H., Snitzer, J. B., & Bridge, J. A. (1982b). Enhancement of recovery of *Legionella pneumophila* from contaminated respiratory tract specimens by heat. *J. Clin. Microbiol.*, **16**:1061–5.
Edelstein, P. H., Bibb, W. F., Gorman, G. W., *et al.* (1984a). *Legionella pneumophila* serogroup 9: a cause of human pneumonia. *Ann. Intern. Med.*, **101**:196–8.
Edelstein, P. H., Calarco, K., & Yasui, V. K. (1984b). Antimicrobiol therapy of experimentally induced Legionnaires' disease in guinea pigs. *Am. Rev. Respir. Dis.*, **130**:849–56.
Edelstein, P. H., Beer, K. B., Sturge, J. C., Watson, A. J., & Goldstein, L. C. (1985). Clinical utility of a monoclonal direct fluorescent reagent specific for *Legionella pneumophila*: comparative study with other reagents. *J. Clin. Microbiol.*, **22**:419–21.
Edelstein, P. H., Bryan, R. N., Enns, R. K., Kohne, D. E., & Kacian, D. L. (1987). Retrospective study of Gen-Probe rapid diagnostic system for detection of legionellae in frozen clinical respiratory tract samples. *J. Clin. Microbiol.*, **25**:1022–6.
Edson, D. C., Stiefel, H. E., Wentworth, B. B., & Wilson, D. L. (1979). Prevalence of antibodies to Legionnaires' disease. A seroepidemiologic survey of Michigan residents using the haemagglutination test. *Ann. Intern. Med.*, **90**:691–3.
Eisenstein, B. I., & Engleberg, N. C. (1986). Applied molecular genetics: new tools for microbiologists and clinicians. *J. Infect. Dis.*, **153**:416–30.
Elder, E. M., Brown, A., Remington, J. S., Shonnard, J., & Naot, Y. (1983). Microenzyme-linked immunosorbent assay for detection of immunoglobulin G and immunoglobulin M antibodies to *Legionella pneumophila*. *J. Clin. Microbiol.*, **17**:112–21.
England, A. C., McKinney, R. M., Skaliy, P., & Gorman, G. W. (1980). A fifth serogroup of *Legionella pneumophila*. *Ann. Intern. Med.*, **93**:58–9.
Engleberg, N. C., Carter, C., Demarsh, P., Drutz, D. J., & Eisenstein, B. I. (1986). A *Legionella*-specific DNA probe detects organisms in lung tissue homogenates from intranasally inoculated mice. *Isr. J. Med. Sci.*, **22**:703–5.
Fallon, R. J. (1986). Identification of *Legionella pneumophila* with commercially available immunofluorescence test. *J. Clin. Pathol.*, **39**:693–4.
Fallon, R. J., & Abraham, W. H. (1983). Experience with heat-killed antigens of *L.longbeachae* serogroups 1 and 2, and *L.jordanis* in the indirect fluorescent antibody test. *Zentralbl. Bakteriol. Mikrobiol. Hyg. (A)*, **255**:8–14.
Farrington, M., & French, G. L. (1983). *Legionella pneumophila* seen in Gram stains of respiratory secretions and recovered from conventional blood cultures. *J. Infect.*, **6**:123–7.
Farshy, C. E., Klein, G. C., & Feeley, J. C. (1978). Detection of antibodies to Legionnaires' disease organism by microagglutination and micro-enzyme-linked immunosorbent assay tests. *J. Clin. Microbiol.*, **7**:327–31.

Feeley, J. C., Gorman, G. W., Weaver, R. E., Mackal, D. C., & Smith, H. W. (1978). Primary isolation media for the Legionnaires' disease bacterium. *J. Clin. Microbiol.*, **8**:320–5.
Feeley, J. C., Gibson, R. J., Gorman, G. W., *et al.* (1979). Charcoal-yeast extract agar: primary isolation medium for *Legionella pneumophila*. *J. Clin. Microbiol.*, **10**:437–41.
Feltham, R. K. A., Power, A. K., Pell, P. A., & Sneath, P. H. A. (1978). A simple method for storage of bacteria at −76 °C. *J. Appl. Bacteriol.*, **44**:313–16.
Fey, H., & Suter, M. (1979). A novel method for the production of *Salmonella* flagellar antigen used for the preparation of H antisera. *Zentralbl. Bakteriol. Mikrobiol. Hyg. (A)*, **243**:216–25.
Finnerty, W. R., Makula, R. A., & Feeley, J. C. (1979). Cellular lipids of the Legionnaires' disease bacterium. *Ann. Intern. Med.*, **90**:631–4.
Fisher-Hoch, S., Hudson, M. J., & Thompson, M. H. (1979). Identification of a clinical isolate as *Legionella pneumophila* by gas chromatography and mass spectrometry of cellular fatty acids. *Lancet*, **ii**:323–5.
Fliermans, C. B., Cherry, W. B., Orrison, L. H., Smith, S. J., Tison, D. L., & Pope, D. H. (1981). Ecological distribution of *Legionella pneumophila*. *Appl. Environ. Microbiol.*, **41**:9–16.
Fraser, D. W., Tsai, T. R., Orenstein, W., *et al.* (1977). Legionnaires' disease: description of an epidemic of pneumonia. *N. Engl. J. Med.*, **297**:1189–97.
Garrity, G. M., Brown, A., & Vickers, R. M. (1980). *Tatlockia* and *Fluoribacter*: two new genera of organisms resembling *Legionella pneumophila*. *Int. J. Sys. Bacteriol.*, **30**:609–14.
Gaultney, J. B., Wende, R. D., & Williams, R. P. (1971). Microagglutination procedures for febrile agglutination tests. *Appl. Microbiol.*, **22**:635–40.
Gilbart, J. (1985). Chromatographic identification and detection of the *Legionella* species. PhD Thesis, University of London.
Gilbart, J., & Collins, M. D. (1985). High-performance liquid chromatographic analysis of ubiquinones from new *Legionella* species. *FEMS Microbiol. Lett.*, **26**:77–82.
Gimenez, F. (1964). Staining rickettsiae in yolk-sac cultures. *Stain Technol.*, **39**:135–40.
Glick, T. H., Gregg, M. B., Berman, B., Mallison, G., Rhodes, W. W., & Kassanoff, I. (1978). Pontiac fever: An epidemic of unknown etiology in a health department: I. clinical and epidemiological aspects. *Am. J. Epidemiol.*, **107**:149–60.
Gorman, G. W., Feeley, J. C., Steigerwalt, A. G., Edelstein, P. H., Moss, C. W., & Brenner, D. J. (1985). *Legionella anisa*: a new species of *Legionella* isolated from potable waters and a cooling tower. *Appl. Environ. Microbiol.*, **49**:305–9.
Gosting, L. H., Cabrian, K., Sturge, J. C., & Goldstein, L. C. (1984). Identification of a species-specific antigen in *Legionella pneumophila* by a monoclonal antibody. *J. Clin. Microbiol.*, **20**:1031–5.
Greaves, P. W. (1980). New methods for the isolation of *Legionella pneumophila*. *J. Clin. Pathol.*, **33**:581–4.
Grimont, P. A. D., Grimont, F., Desplaces, N., & Tchen, P. (1985). DNA probe specific for *Legionella pneumophila*. *J. Clin. Microbiol.*, **21**:431–7.
Grob, R. L. (1985). *Modern Practice of Gas Chromatography*, 2nd edn. London: Wiley.
Grob, K., & Grob, K. (1981). Splitless injection and the solvent effect. In: Bertsch, W., Jennings, W., & Kaiser, R. E. (eds). *Recent Advances in Capillary Gas Chromotography*. Heidelberg: Huthig, pp.455–74.
Guillet, J. G., Hoebeke, J., Tram, C., Marullo, S., & Strosberg, A. D. (1983). Characterization, serological specificity and diagnostic possibilities of monoclonal antibodies against *Legionella pneumophila*. *J. Clin. Microbiol.*, **18**, 793–7.
Halper, L. A., & Norton, S. J. (1975). Regulation of cyclopropane fatty acid biosynthesis

by variation in enzyme activities. *Biochem. Biophys. Res. Commun.*, **62**:683–8.
Harrison, T. G. (1984). The serological diagnosis of Legionnaires' disease. PhD. Thesis, University of London.
Harrison, T. G., & Taylor, A. G., (1982a). Diagnosis of *Legionella pneumophila* infections by means of formolised yolk sac antigens. *J. Clin. Pathol.*, **35**:211–14.
Harrison, T. G., & Taylor, A. G. (1982b). A rapid microagglutination test for the diagnosis of *Legionella pneumophila* (serogroup 1) infection. *J. Clin. Pathol.*, **35**:1028–31.
Harrison, T. G., Dournon, E., & Taylor, A. G. (1987). Evaluation of sensitivity of two serological tests for diagnosing pneumonia caused by *Legionella pneumophila* serogroup 1. *J. Clin. Pathol.*, **40**:77–82.
Harvey, D. J. (1982). Picolinyl esters as derivatives for the structural determination of long chain branched and unbranched fatty acids. *Biomed. Mass Spectrom.*, **9**:33–8.
Hébert, G. A. (1981). Hippurate hydrolysis by *Legionella pneumophila*. *J. Clin. Microbiol.*, **13**:240–2.
Hébert, G. A., Pittman, B., McKinney, R. M., & Cherry, W. B. (1972). *The Preparation and Physiochemical Characterization of Fluorescent Antibody Reagents.* Atlanta, Ga.: US Department of Health, Education and Welfare, Centers for Disease Control.
Hébert, G. A., Steigerwalt, A. G., & Brenner, D. J. (1980). *Legionella micdadei* species nova: classification of a third species of *Legionella* associated with human pneumonia. *Curr. Microbiol.*, **3**:255–7.
Helms, C. M., Johnson, W., Renner, E. D., Hierholzer, W. J., Wintermeyer, L. A., & Viner, J. P. (1980). Background prevalence of microagglutination antibodies to *Legionella* pneumophila serogroups 1, 2, 3, and 4. *Infect. Immun.*, **30**:612–14.
Herbrink, P., Meenhorst, P. L., Groothuis, D. G., *et al.* (1983). Detection of antibodies against *Legionella pneumophila* serogroups 1 to 6 and the Leiden-1 strain by micro ELISA and immunofluorescence assay. *J. Clin. Pathol.*, **36**:1246–52.
Hernandez, J. F., Delattre, J. M., & Oger, C. (1983). Thermoresistance des *Legionella*. *Ann. Microbiol.*, **134b**:421–7.
Herwaldt, L. A., Gorman, G. W., McGrath, T., *et al.* (1984). A new *Legionella* species, *Legionella feeleii* species nova, causes Pontiac fever in an automobile plant. *Ann. Intern. Med.*, **100**:333–8.
Holliday, M. G. (1980). The diagnosis of Legionnaires' disease by counterimmunoelectrophoresis. *J. Clin. Pathol.*, **33**:1174–8.
Horwitz, M. A., & Silverstein, S. C. (1980). Legionnaires' disease bacterium (*Legionella pneumophila*) multiplies intracellularly in human monocytes. *J. Clin. Invest.*, **66**:441–50.
Horwitz, M. A., & Silverstein, S. C. (1983). Intracellular multiplication of Legionnaires' disease bacteria (*Legionella pneumophila*) in human monocytes is reversibly inhibited by erythromycin and rifampin. *J. Clin. Invest.*, **71**:15–26.
Huang, C., Huang, H. S., Glembourtt, M., Liu, C., & Cohen, S. N. (1985). Monoclonal antibody specific for double-stranded DNA: a non-radioactive probe method for detection of DNA hybridization. In: Kingsbury, D. T., & Falkow, S. (eds), *Rapid Detection and Identification of Infectious Agents.* London: Academic Press, pp.257–70.
Jantzen, E., & Bryn, K. (1985). Whole-cell and lipopolysaccharide fatty acids and sugars of gram-negative bacteria. In: Goodfellow, M., & Minnikin, D. E. (eds), *Chemical Methods in Bacterial Systematics.* London: Academic Press, pp.145–71.
Jantzen, E., Bryn, K., Hagen, N., Bergan, T., & Bovre, K. (1978). Fatty acids and monosaccharides of *Neisseriaceae* in relation to established taxonomy. *NIPH Ann.*, **2**:59–71.
Jennings, W. (1980). Gas-chromatography with glass capillary columns, 2nd edn. London: Academic Press.

Jennings, W. (1981). *Comparisons of Fused Silica and other Glass Capillary Columns*. Heidelberg: Huthig.

Johnson, D. A., Wagner, K. F., Blanks, J., & Salter, J. (1985). False-positive direct fluorescent antibody testing for *Legionella*. *JAMA*, **253**:40–1.

Joly, J. R., & Ramsay, D. (1985). Use of monoclonal antibodies in the diagnosis and epidemiologic studies of legionellosis. *Clin. Lab. Med.*, 5:561–74.

Joly, J. R., Chen, Y., & Ramsay, D. (1983). Serogrouping and subtyping of *Legionella pneumophila* with monoclonal antibodies. *J. Clin. Microbiol.*, **18**:1040–6.

Joly, J. R., Dery, P., Gauvreau, L., Cote, L., & Trepanier, C. (1986a). Legionnaires' disease caused by *Legionella dumoffii* in distilled water. *Can. Med. Assoc. J.*, **135**:1274–7.

Joly, J. R., McKinney, R. M., Tobin, J. O.' H., Bibb, W. F., Watkins, I. D., & Ramsay, D. (1986b). Development of a standardized subgrouping scheme for *Legionella pneumophila* serogroup 1 using monoclonal antibodies. *J. Clin. Microbiol.*, **23**:768–71.

Kalweit, W. H., Winn, W. C., Rocco, T. A., & Girod, J. C. (1982). Hemodialysis fistula infections caused by *Legionella pneumophila*. *Ann. Intern. Med.*, **96**:173–5.

Karr, D. E., Bibb, W. F., & Moss, C. W. (1982). Isoprenoid quinones of the genus *Legionella*. *J. Clin. Microbiol.*, **15**:1044–8.

Kirby, B. D., Snyder, K. M., Meyer, R. D., & Finegold, S. M. (1980). Legionnaires' disease: report of sixty-five nosocomially acquired cases and review of the literature. *Medicine (Baltimore)*, **59**:188–205.

Kirsop, B. E., & Snell, J. J. S. (eds). (1984). *Maintenance of Microorganisms: a Manual of Laboratory Methods*. London: Academic Press.

Kleger, B., & Hartwig, R. A. (1979). Development and evaluation of a card agglutinin titer (cat) test for serodiagnosis of Legionnaires' disease. *Publ. Health Lab.*, **38**:247–54.

Knox, J. H. (1981). *High-performance Liquid Chromatography*. Edinburgh: Edinburgh University Press.

Kohler, G., & Milstein, C. (1975). Continuous cultures of fused cells secreting antibody of predefined specificity. *Nature*, **256**: 495–7.

Kohler, R. B., & Sathapatayavongs, B. (1983). Recent advances in the diagnosis of serogroup 1. *L.pneumophila* pneumonia by detection of urinary antigen. *Zentralbl. Bakteriol. Mikrobiol. Hyg. (A)*, **255**:102–7.

Kohler, R. B., Zimmerman, S. E., Wilson, E., *et al.* (1981). Rapid radioimmunoassay diagnosis of Legionnaires' disease: detection and partial characterization of urinary antigen. *Ann. Intern. Med.*, **94**:601–5.

Kohler, R. B., Winn, W. C., Girod, J. C., & Wheat, L. J. (1982). Rapid diagnosis of pneumonia due to *Legionella pneumophila* Serogroup 1. *J. Infect. Dis.*, **146**:444.

Kohler, R. B., Winn, W. C., & Wheat, L. J. (1984). Onset and duration of urinary antigen excretion in Legionnaires' disease. *J. Clin. Microbiol.*, **20**:605–7.

Kohne, D. E., Steigerwalt, A. G., & Brenner, D. J. (1984). Nucleic acid probe specific for members of the genus *Legionella*. In: Thornsberry, C., Balows, A., Feeley, J. C., & Jakubowski, W. (eds), *Legionella: Proceedings of the 2nd International Symposium*. Washington DC: Am. Soc. Microbiol., pp.107–8.

Krivankova, L., & Dadak, V. (1980). Semimicro extraction of ubiquinone and menaquinone from bacteria. In: McCormick, D. B., & Wright, L. D. (eds), *Methods in Enzymology*, vol. 67, pp.111–14.

Lambert, M. A., & Moss, C. W. (1983). Comparison of the effects of acid and base hydrolyses on hydroxy and cyclopropane fatty acids in bacteria. *J. Clin. Microbiol.*, **18**:1370–7.

Langer, P. R., Waldrop, A. A., & Ward, D. C. (1981). Enzymatic synthesis of biotin-labelled polynucleotides: novel nucleic acid affinity probes. *Proc. Natl. Acad. Sci. USA*, **78**:6633–7.

Larsson, L., & Odham, G. (1984). Injection principles in capillary gas chromatographic analysis of bacterial fatty acids. *J. Microbiol. Methods*, **3**:77–82.

Lennette, D. A., Lennette, E. T., Wentworth, B. B., French, M. L. V., & Lattimer, G. L. (1979). Serology of Legionnaires' disease: comparison of indirect fluorescent antibody, immune adherence hemagglutination, and indirect hemagglutination tests. *J. Clin. Microbiol.*, **10**:876–9.

Lim, C. K. (1986). *HPLC of Small Molecules: a Practical Approach*. Oxford: IRL Press.

Lind, K., Collins, M. T., & Aalund, O. (1983). *Legionella* microagglutination test. A serological study of Danish patients. *Zentralbl. Bakteriol. Mikrobiol. Hyg. (A)*, **225**:115–19.

Liu, F., & Wright, D. N. (1984). Gram stain in Legionnaires' disease. *Am. J. Med.*, **77**:549–50.

Macfarlane, J. T., Finch, R. G., Ward, M. J., & Macrae, A. D. (1982). Hospital study of adult community-acquired pneumonia. *Lancet*, **ii**:255–8.

Macleod, K. B., Patton, C. M., Klein, G. G., & Feeley, J. C. (1984). Growth of *Francisella tularensis* on media for the cultivation of *Legionella pneumophila*. In: Thornsberry, C., Balows, A., Feeley, J. C., & Jakubowski, W. (eds), *Legionella: Proceedings of the 2nd International Symposium*. Washington DC: Am. Soc. Microbiol., pp.20–1.

Macrae, A. D., & Lewis, M. J. (1977). Legionnaires' disease in Nottingham. *Lancet*, **ii**:1225–6.

Macrae, A. D., Greaves, P. W., & Platts, P. (1979). Isolation of *Legionella pneumophila* from blood culture. *Br. Med. J.*, 1189–90.

Mangiafico, J. A., Hedlund, K. W., & Knott, A. R. (1981). Rapid and sensitive method for quantitation of *Legionella pneumophila* Serogroup 1 antigen from human urine. *J. Clin. Microbiol.*, **13**:843–5.

Maniatis, T., Fritsch, E. F., & Sambrook, J. (1982). *Molecular cloning: a Laboratory Manual*. New York: Coldspring Harbor Laboratory, pp.320–6.

Mayaud, C., Carette, M. F., Dournon, E., Buré, A., Francois, T., & Akoun, G. (1984). Clinical features and prognosis of severe pneumonia caused by *Legionella pneumophila*. In: Thornsberry, C., Balows, A., Feeley, J. C., & Jakubowski, W. (eds), *Legionella: Proceedings of the 2nd International Symposium*. Washington DC: Am. Soc. Microbiol., pp.11–12.

Mayberry, W. R. (1981). Dihydroxy and monohydroxy fatty acids in *Legionella pneumophila*. *J. Bacteriol.*, **147**:373–81.

Mayberry, W. R. (1984). Composition and structure of dihydroxy, monohydroxy and unsaturated nonhydroxy fatty acids of *Legionella* species. In: Thornsberry, C., Balows, A., Feeley, J. C., & Jakubowski, W. (eds), *Legionella: Proceedings of the 2nd International Symposium*. Washington DC: Am. Soc. Microbiol., pp.79–81.

Mayock, R., Skale, B., & Kohler, R. B. (1983). *Legionella pneumophila* pericarditis proved by culture of pericardial fluid. *Am. J. Med.*, **75**:534–6.

McCabe, R. E., Baldwin, J. C., McGregor, C. A., Miller, D. C., & Vosti, K. L. (1984). Prosthetic valve endocarditis caused by *Legionella pneumophila*. *Ann. Intern. Med.*, **100**:525–7.

McDade, J. E., Shepard, C. C., Fraser, D. W., *et al.* (1977). Legionnaires' disease: isolation of a bacterium and domonstration of its role in respiratory disease. *N. Engl. J. Med.*, **297**:1197–203.

McDade, J. E., Brenner, D. J., & Bozeman, F. M. (1979). Legionnaires' disease bacterium isolated in 1947. *Ann. Intern. Med.*, **90**:659–61

McKinney, R. M. (1985). Diagnosis: direct immunofluorescent antibody test. In: Katz, S. M. (ed.), *Legionellosis*, vol. 2. Bocca Raton: CRC Press, pp.23–32.

McKinney, R. M., Thacker, L., Harris, P. P., *et al.* (1979a). Four serogroups of Legionnaires' disease bacteria defined by direct immunofluorescence. *Ann. Intern. Med.*, **90**:621–4.

McKinney, R. M., Thomason, B. M., Harris, P. P., *et al.* (1979b). Recognition of a new serogroup of Legionnaires' disease bacterium. *J. Clin. Microbiol.*, **9**:103–7.
McKinney, R. M., Wilkinson, H. W., Sommers, H. M., *et al.* (1980). *Legionella pneumophila* serogroup six: isolation from cases of legionellosis, identification by immunofluorescence staining, and immunological response to infection. *J. Clin. Microbiol.*, **12**:395–401.
McKinney, R. M., Porschen, R. K., Edelstein, P. H., *et al.* (1981). *Legionella longbeachae* species nova, another etiologic agent of human pneumonia. *Ann. Intern. Med.*, **94**:739–43.
McKinney, R. M., Thacker, L., Wells, D. E., Wong, M. C., Jones, W. J., & Bibb, W. F. (1983). Monoclonal antibodies to *Legionella pneumophila* serogroup 1: possible applications in diagnostic tests and epidemiologic studies. *Zentralbl. Bakteriol. Mikrobiol. Hyg. (A)*, **255**:91–5.
Meenhorst, P. L., Reingold, A. L., Groothuis, D. G., *et al.* (1985). Water-related nosocomial pneumonia caused by *Legionella pneumophila* serogroups 1 and 10. *J. Infect. Dis.*, **152**:356–64.
Melton, D. A., Krieg, P. A., Rebagliati, M. R., Maniatis, T., Zinn, K., & Green, M. R. (1984). Efficient *in vitro* synthesis of biologically active RNA and RNA hybridization probes from plasmids containing a bacteriophage SP6 promotor. *Nucleic Acids Res.*, **12**:7035–56.
Meyer, R. D. (1984). Legionnaires' disease: aspects of nosocomial infection. *Am. J. Med.*, **76**:657–63.
Meyer, R. D., Edelstein, P. H., Kirby, B. D., *et al.* (1980). Legionnaires' disease: unusual clinical and laboratory features. *Ann. Intern. Med.*, **93**:240–3.
Miller, L. T. (1982). Single derivatization method for routine analysis of bacterial whole-cell fatty acid methyl esters, including hydroxy acids. *J. Clin. Microbiol.*, **16**:584–6.
Minnikin, D. E., O'Donnell, A. G., Goodfellow, M., *et al.* (1984). An integrated procedure for the extraction of bacterial isoprenoid quinones and polar lipids. *J. Microbiol. Methods*, **2**:233–41.
Miyazaki, T., Koga, H., Nakashima, M., *et al.* (1985). Production of monoclonal antibodies against *Legionella pneumophila* serogroup 1. *Microbiol. Immunol.*, **29**:275–84.
Morris, G. K., Patton, C. M., Feeley, J. C., *et al.* (1979). Isolation of the Legionnaires' disease bacterium from environmental samples. *Ann. Intern. Med.*, **90**:664–6.
Morris, G. K., Steigerwalt, A. G., Feeley, J. C., *et al.* (1980). *Legionella gormanii* sp. nov. *J. Clin. Microbiol.*, **12**:718–21.
Moss, C. W., & Dees, S. B. (1979a). Further studies of the cellular fatty acid composition of Legionnaires' disease bacteria. *J. Clin. Microbiol.*, **9**:648–9.
Moss, C. W., & Dees, S. B. (1979b). Cellular fatty acid composition of WIGA, a Rickettsia-like agent similar to the Legionnaires' disease bacterium. *J. Clin. Microbiol.*, **10**:390–1.
Moss, C. W., & Guerrant, G. O. (1983). Separation of bacterial ubiquinones by reverse-phase high-pressure liquid chromatography. *J. Clin. Microbiol.*, **18**:15–17.
Moss, C. W., Lambert, M. A., Merwin, W. H. (1974). Comparison of rapid methods for analysis of bacterial fatty acids. *Appl. Microbiol.*, **28**:80–5.
Moss, C. W., Weaver, R. E., Dees, S. B., & Cherry, W. B. (1977). Cellular fatty acid composition of isolates from Legionnaires' disease. *J. Clin. Microbiol.*, **6**:140–3.
Moss, C. W., Karr, D. E., & Dees, S. B. (1981). Cellular fatty acid composition of *Legionella longbeachae* sp. nov. *J. Clin. Microbiol.*, **14**:692–4.
Moss, C. W., Bibb, W. F., Karr, D. E., Guerrant, G. O., & Lambert, M. A. (1983). Cellular fatty acid composition and ubiquinone content of *Legionella feeleii* sp. nov. *J. Clin. Microbiol.*, **18**:917–19.

Moyer, N. P., Lange, A. F., Hall, N. H., & Hausler, W. J. (1984). Application of capillary gas-liquid chromatography for subtyping *Legionella pneumophila*. In Thornsberry, C., Balows, A., Feeley, J. C., & Jakubowski, W. (eds), *Legionella: Proceedings of the 2nd International Symposium*. Washington DC: Am. Soc. Microbiol, pp.274–6.

Muller, H. E. (1981). The thermic stability of *Legionella pneumophila*. *Zentralb. Bakteriol. Mikrobiol. Hyg. (B)*, **172**:524–7.

Nagington, J., Wreghitt, T. G., & Smith, D. J. (1979a). How many Legionnaires? *Lancet*, **i**:536–7.

Nagington, J., Wreghitt, T. G., Tobin, J. O.' H., & Macrae, A. D. (1979b). The antibody response in Legionnaires' disease. *J. Hyg. (Camb.)*, **83**:377–81.

Odham, G., Larsson, L., Mardh, O. K. (eds) (1984). *Gas-chromatograpy/mass spectrometry: Applications in Microbiology*. New York: Plenum Press.

Orrison, L. H., Bibb, W. F., Cherry, W. B., & Thacker, L. (1983a). Determination of antigenic relationships among legionellae and non-legionellae by direct fluorescent-antibody and immunodiffusion tests. *J. Clin. Microbiol.*, **17**:332–7.

Orrison, L. H., Cherry, W. B., Tyndall, R. L., *et al.* (1983b). *Legionella oakridgensis*; unusual new species isolated from cooling tower water. *Appl. Environ. Microbiol.*, **45**:536–45.

Para, M. F., & Plouffe, J. F. (1983). Production of monoclonal antibodies to *Legionella pneumophila* serogroups 1 and 6. *J. Clin. Microbiol.*, **18**:895–900.

Pasculle, A. W., Feeley, J. C., Gibson, R. J., *et al.* (1980). Pittsburgh pneumonia agent: direct isolation from human lung tissue. *J. Infect. Dis.*, **141**:727–32.

Petitjean, F., Rajagopalan, P., Hoebeke, J., & Dournon, E. (1986). Correlation of virulence of *L.pneumophila* serogroup 1 and monoclonal binding. *Program and abstracts of the 26th Interscience Conference on Antimicrobial Agents and Chemotherapy*. Washington DC: Am. Soc. Microbiol., pp.293.

Pine, L., Hoffman, P. S., Malcolm, G. B., Benson, R. F., & Gorman, G. W. (1984). Whole-cell peroxidase test for identification of *Legionella pneumophila*. *J. Clin. Microbiol.*, **19**:286–90.

Plouffe, J. F., Para, M. F., Maher, W. E., Hackman, B., & Webster, L. (1983). Subtypes of *Legionella pneumophila* serogroup 1 associated with different attack rates. *Lancet*, **ii**:649–50.

Rajagopalan, P., & Dournon, E. (1986). A simple method for direct isolation of *Legionella pneumophila* from blood. *Isr. J. Med. Sci.*, **22**:757.

Renz, M., & Kurz, C. (1984). A colorimetric method for DNA hydridization. *Nucleic Acids Res.*, **12**:3435–44.

Rietschel, E. T., Gottert, H., Luderitz, O., & Westphal, O. (1972). Nature and linkages of the fatty acids present in the Lipid-A component of salmonella lipopolysaccharide. *Eur. J. Biochem.*, **28**:166–73.

Rihs, J. D., Yu, V. L., Zuravleff, J. J., Goetz, A., & Muder, R. R. (1985). Isolation of *Legionella pneumophila* from blood with the BACTEC system: a prospective study yielding positive results. *J. Clin. Microbiol.*, 22:422–4.

Rodgers, F. G., & Laverick, A. (1984). *Legionella pneumophila* serogroup 1 flagellar antigen in a passive hemagglutination test to detect antibodies to other *Legionella* species. In: Thornsberry, C., Balows, A., Feeley, J. C., & Jakubowski, W. (eds), *Legionella*: *Proceedings of the 2nd International Symposium*. Washington DC: Am. Soc. Microbiol., pp.42–4

Rudin, J. E., Wing, E. J., & Yee, R. B. (1984). An ongoing outbreak of *Legionella micdadei*. In: Thornsberry, C., Balows, A., Feeley, J. C., Jakubowski, W. (eds), *Legionella*: *Proceedings of the 2nd International Symposium*. Washington DC: Am. Soc. Microbiol., pp.227–9.

Ryhage, R., & Stenhagen, E. (1960). Mass spectrometry in lipid research. *J. Lipid Research*, **1**:361–90.

Sampson, J. S., Plikaytis, B. B., & Wilkinson, H. W. (1986). Immunologic response of patients with legionellosis against major protein-containing antigens of *Legionella pneumophila* serogroup 1 as shown by immunoblot analysis. *J. Clin. Microbiol.*, **23**:92–9.

Samuel, D., Harrison, T. G., Carr, J., & Taylor, A. G. (1987). Serological response in Legionnaires' disease analysed by immunoblotting. In: Fleurette, J., Bornstein, N., Marmet, D., & Surgot, M. (eds), Colloque Legionella, 6–7 May 1987. Lyon: Fondation Marcel Merieux, pp.69–74.

Sathapatayavongs, B., Kohler, R. B., Wheat, L. J., *et al.* (1982). Rapid diagnosis of Legionnaires' disease by urinary antigen detection: comparison of ELISA and radioimmunoassay. *Am. J. Med.*, **72**:576–82.

Sathapatayavongs, B., Kohler, R. B., Wheat, L. J., White, A., & Winn, W. C. (1983). Rapid diagnosis of Legionnaires' disease by latex agglutination. *Am. Rev. Respir. Dis.*, **127**:559–62.

Saunders, N. A., Kachwalla, N., Harrison, T. G., & Taylor, A. G. (1988). Cloned nucleic acid probes for detection and identification of legionellae. *Analyt. Proc.* **25**.

Schlanger, G., Lutwick, L. I., Kurzman, M., Hoch, B., & Chandler, F. W. (1984). Sinusitis caused by *Legionella pneumophila* in a patient with the acquired immune deficiency syndrome. *Am. J. Med.*, **77**:957–60.

Selander, R. K., McKinney, R. M., Whittam, T. S., *et al.* (1985). Genetic structure of populations of *Legionella pneumophila*. *J. Bacteriol.*, **163**:1021–37.

Sethi, K. K., Drueke, V., & Brandis, H. (1983). Hybridoma-derived monoclonal immunoglobulin M antibodies to *Legionella pneumophila* serogroup 1 with diagnostic potential. *J. Clin. Microbiol.*, **17**:953–7.

Shearer, M. J. (1986). Assay of K vitamins in tissues by high-performance liquid chromatography with special reference to ultraviolet detection. In: Chytil, F., & McCormick, D. B. (eds), *Methods in Enzymology*, vol.123, pp.235–51.

Sillis, M., & Andrews, B. E. (1978). A simple test for *Mycoplasma pneumoniae* IgM. *Zentralbl. Bakteriol. Mikrobiol. Hyg. (A)*, **241**:239–40.

Snyder, L. R., & Kirkland, J. J. (1979). *Introduction to Modern Liquid Chromatography*, 2nd edn. New York: Wiley.

Soriano, F., Aguilar, L., & Garces, J. L. (1982). Simple immunodiffusion test for detecting antibodies against *Legionella pneumophila* serotype 1. *J. Clin. Microbiol.*, **15**:330–1.

Southern, E. M. (1975). Detection of specific sequences among DNA fragments separated by gel electrophoresis. *J. Mol. Biol.*, **98**:503–17.

Stout, J., Yu, V. L., Vickers, R. M., & Shonnard, J. (1982). Potable water supply as the hospital reservoir for Pittsburgh pneumonia agent. *Lancet*, **i**:471–2.

Tamaoka, J. (1986). Analysis of bacterial menaquinone mixtures by reverse-phase high-performance liquid chromatography. In: Chytil, F., & McCormick, D. B. (eds), *Methods in Enzymology*, vol.123, pp.251–6.

Tang, P. W., deSavigny, D., & Toma, S. (1982). Detection of *Legionella* antigenuria by reverse passive agglutination. *J. Clin. Microbiol.*, **15**:998–1000.

Tang, P.W., Toma, S., Moss, C. W., Steigerwalt, A. G., Cooligan, T. G., & Brenner, D. J. (1984). *Legionella bozemanii* serogroup 2: a new etiological agent. *J. Clin. Microbiol.*, **19**:30–3.

Taylor, A. G., & Harrison, T. G. (1979). Legionnaires' disease caused by *Legionella pneumophila* serogroup 3. *Lancet*, **i**:47.

Taylor, A. G., & Harrison, T. G. (1983). Serological tests for *Legionella pneumophila* serogroup 1 infections. *Zentralbl. Bakteriol. Mikrobiol. Hyg. (A)*, **255**:20–6.

Taylor, A. G., Harrison, T. G., Dighero, M. W. & Bradstreet, C. M. P. (1979). False positive reactions in the indirect immunofluorescent antibody test for Legionnaires' disease eliminated by use of formolised yolk-sac antigen. *Ann. Intern. Med.*, **90**:686–9.

Temperanza, A. M., Di Capua, A., Ciarrocchi, S., Ciceroni, L., & Castellani Pastoris, M. (1986). More experience on the microagglutination test in the diagnosis of *Legionella pneumophila* infection. *Microbiologica*, **9**:71–9.

Tenover, F. C., Carlson, L., Goldstein, L., Sturge, J., & Plorde, J. J. (1985). Confirmation of *Legionella pneumophila* cultures with a fluorescein-labelled monoclonal antibody. *J. Clin. Microbiol.*, **21**:983–4.

Tenover, F. C., Edelstein, P. H., Goldstein, L. C., Sturge, J. C., & Plorde, J. J. 91986). Comparison of cross-staining reactions by *Pseudomonas* spp. and fluorescein-labelled polyclonal and monoclonal antibodies directed against *Legionella pneumophila*. *J. Clin. Microbiol.*, **23**:647–9.

Thacker, L., McKinney, R. M., Moss, C. W., Sommers, H. M., Spivack, M. L., & O'Brien, T. F. (1981). Thermophilic sporeforming bacilli that mimic fastidious growth characteristics and colonial morphology of legionellae. *J. Clin. Microbiol.*, **13**:794–7.

Thacker, W. L., Wilkinson, H. W., & Benson, R. F. (1983). Comparison of slide agglutination test and direct immunofluorescence assay for identification of *Legionella* isolates. *J. Clin. Microbiol.*, **18**:1113–18.

Thacker, W. L., Plikaytis, B. B., & Wilkinson, H. W. (1985a). Identification of 22 *Legionella* species and 33 serogroups with the slide agglutination test. *J. Clin. Microbiol.*,**21**:779–82.

Thacker, W. L., Wilkinson, H. W., Plikaytis, B. B., *et al.* (1985b). Second serogroup of *Legionella feeleii* strains isolated from humans. *J. Clin. Microbiol.*, **22**:1–4.

Thacker, W. L., Benson, R. F., Wilkinson, H. W., *et al.* (1986). 11th serogroup of *Legionella pneumophila* isolated from a patient with fatal pneumonia. *J. Clin. Microbiol.*, **23**:1146–7.

Thacker, W. L., Wilkinson, H. W., Benson, R. F., & Brenner, D. J. (1987). *Legionella pneumophila* serogroup 12 isolated from human and environmental sources. *J. Clin. Microbiol.*, **25**:569–70.

Theaker, J. M., Tobin, J. O.' H., Jones, S. E. C., Kirkpatrick, P., Vina, M. I. & Fleming, K. A. (1987). Immunohistological detection of *Legionella pneumophila* in lung sections. *J. Clin. Pathol.*, **40**:143–6.

Tiemeier, D., Engquist, L., & Leder, P. (1976). Improved derivative of a phage EK2 vector for cloning recombinant DNA. *Nature*, **263**:526–7.

Tilton, R. C. (1979). Legionnaires' disease antigen detected by enzyme-linked immunosorbent assay. *Ann. Intern. Med.*, **90**:697–8.

Tison, D. L., Baross, J. A., & Seidler, R. J. (1983). *Legionella* in aquatic habitats in the Mount Saint Helens blast zone. *Curr. Microbiol.*, **9**:345–8.

Tobin, J. O.' H., Beare, J., Dunhill, M. S., *et al.* (1980). Legionnaires' disease in a transplant unit: isolation of the causative agent from shower baths. *Lancet*, **ii**:118–21.

Tsai, T. F., & Fraser, D. W. (1978). The diagnosis of Legionnaires' disease. *Ann. Intern. Med.*, **89**:413–14.

Turner, L. H. (1968). Leptospirosis II: Serology. *Trans. R. Soc. Trop. Med. Hyg.*, **62**:880–99.

van Ketel, R. J., ter Schegget, J., & Zanen, H. C. (1984). Molecular epidemiology of *Legionella pneumophila* serogroup 1. *J. Clin. Microbiol.*, **20**:362–4.

Vecchio, T. J. (1986). Predictive value of a single diagnostic test in unselected populations. *N. Engl. J. Med.*, **274**:1171–3.

Vickers, R. M., Brown, A., & Garrity, G. M. (1981). Dye-containing buffered charcoal-yeast extract medium for differentiation of members of the family *Legionellaceae*. *J. Clin. Microbiol.*, **13**:380–2.

Vildé, J. L., Dournon, E., & Rajagopalan, P. (1986). Inhibition of *Legionella pneumophila* multiplication within human macrophages by antimicrobial agents. *Antimicrob. Agents Chemother.*, **30**:743–8.

Vulliet, P., Markey, S. P., & Tornabene, T. G. (1974). Identification of methoxyester artefacts produced by methanolic-HCl solvolysis of the cyclopropane fatty acids of the genus *Yersinia*. *Biochem. Biophys. Acta*, **348**:299–301.

Wadowsky, R. M., & Yee, R. B. (1981). Glycine-containing selective medium for isolation of *Legionellaceae* from environmental specimens. *Appl. Environ. Microbiol.*, **42**:768–72.

Wait, R., & Hudson, M. J. (1985). The use of picolinyl esters for the characterization of microbial lipids: application to the unsaturated and cyclopropane fatty acids of *Campylobacter* species. *Lett. Appl. Microbiol.*, **1**:95–9.

Wait, R., Dennis, P. J., & Hudson, M. J. (1986). The use of gas-chromatography/mass spectrometry for the identification and taxonomy of *Legionella* species. *Biochem. Soc. Trans.*, **14**:965–6.

Wang, W. L. L., Blaser, M. J., Cravens, J., & Johnson, M. A. (1979). Growth survival and resitance of the Legionnaires' disease bacterium. *Ann. Intern. Med.*, **90**:614–18.

Watkins, I. D., & Tobin, J. O.' H. (1984). Studies with monoclonal antibodies to *Legionella* species. In: Thornsberry, C., Balows, A., Feeley, J. C., Jakubowski, W. (eds), *Legionella*: *Proceedings of the 2nd International Symposium*. Washington DC: Am. Soc. Microbiol. pp.259–62.

Watkins, I. D., Tobin, J. O.' H., Dennis, P. J., Brown, W., Newnham, R. S., & Kurtz, J. B. (1985). *Legionella pneumophila* serogroup 1 subgrouping by monoclonal antibodies–an epidemiological tool. *J. Hyg. (Camb.)*, **95**:211–16.

West, M., Burdash, N. M., & Freimuth, F. (1977). Simplified silver-plating stain for flagella. *J. Clin. Microbiol.*, **6**:414–19.

Wilkinson, H. W., & Brake, B. J. (1982). Formalin-killed versus heat-killed *Legionella pneumophila* serogroup 1 antigen in the indirect immunofluorescence assay for legionellosis. *J. Clin. Microbiol.*, **16**:979–81.

Wilkinson, H. W., Farshy, C. E., Fikes, B. J., Cruce, D. D., & Yealy, L. P. (1979a). Measure of immunoglobulin G-, M-, and A-specific titers against *Legionella pneumophila* and inhibition of titers against nonspecific, Gram-negative bacterial antigens in the indirect immunofluorescence test for legionellosis. *J. Clin. Microbiol.*, **10**:685–9.

Wilkinson, H. W., Fikes, B. J., & Cruce, D. D. (1979b). Indirect immunofluorescence test for serodiagnosis of Legionnaires' disease: Evidence for serogroup diversity of Legionnaires' disease bacterial antigens and for multiple specificity of human antibodies. *J. Clin. Microbiol.*, **9**:379–83.

Wilkinson, H. W., Reingold, A. L., Brake, B. J., McGiboney, D. L., Gorman, G. W., & Broome, C. V. (1983). Reactivity of serum from patients with suspected legionellosis against 29 antigens of Legionellaceae and *Legionella*-like organisms by indirect immunofluorescence assay. *J. Infect. Dis.*, **147**:23–31.

Wilkinson, H. W., Thacker, W. L., Steigerwalt, A. G., Brenner, D. J., Ampel, N. M., & Wing, E. J. (1985). Second serogroup of *Legionella hackeliae* isolated from a patient with pneumonia. *J. Clin. Microbiol.*, **22**:488–9.

Wilkinson, H. W., Sampson, J. S., & Plikaytis, B. B. (1986). Evaluation of a commercial gene probe for identification of *Legionella* cultures. *J. Clin. Microbiol.*, **23**:217–20.

Wilkinson, H. W., Thacker, W. L., Benson, R. F., *et al.* (1987). *Legionella birminghamensis* sp. nov. Isolated from a cardiac transplant recipient. *J. Clin. Microbiol.*, **25**:2120–2.

Winberley, N., Willey, S., Sullivan, N., & Bartlett, J. G. (1979). Antibacterial properties of lidocaine. *Chest*, **76**:37–40.

Woodhead, M. A., Macfarlane, J. T., McCracken, J. S., Rose, D. H., & Finch, R. G.

(1987). Prospective study of the aetiology and outcome of pneumonia in the community. *Lancet*, **i**:671–4.

Wreghitt, T. G., Nagington, J., & Gray, J. (1982). An ELISA test for the detection of antibodies to *Legionella pneumophila*. *J. Clin. Pathol.*, **35**:657–60.

Zimmerman, S. E., French, M. L., Allen, S. D., Wilson, E., & Kohler, R. B. (1982). Immunoglobulin M antibody titers in the diagnosis of Legionnaires' disease. *J. Clin. Microbiol.*, **16**:1007–11.

Index

(compiled with the assistance of R. Birtles)

Roman type (25) indicates a textual entry, whereas italics (*25*) show a table or figure entry.